**Fundamentals of Construction Claims**

# Fundamentals of Construction Claims

*A 9-Step Guide for General Contractors, Subcontractors, Architects, Engineers, and Owners*

*William J. McConnell JD, MSCE, PE*

*Library of Congress Cataloging-in-Publication Data applied for*

ISBN: 9781119679905

Cover Design: Wiley
Cover Image: © lamontak590623/iStock/Getty Images

SKY10034656_061322

# Contents

# Acknowledgments

The following colleagues of mine served as technical editors for certain sections of the book: Diana Minchella, Jeffrey Katz, Andrew Sargent, and Ryan Phillips.

# 1

## Introduction

Construction is an industry that is filled with risk and uncertainty. First, very few projects are designed alike, and no two project sites are the same in terms of geotechnical conditions and topography, so each project has its own learning curve. Second, humans manage the development, design, administration, and construction of projects and the skill set, personality mix, and experience level of project team members vary from project to project, which affects overall project performance. Third, the construction of a project includes a multitude of companies, including an owner, an architect, multiple engineering firms, third-party inspectors, a general contractor, multiple subcontractors, multiple vendors, utility companies, a variety of jurisdictional agencies, and others. Each entity operates with its own urgency and has a fixed number of resources to allocate to a given project, so these variables influence performance.

When these three factors are considered together, the likelihood that a construction project will not involve disputes amongst certain parties is low, even though the parties to a construction project typically have the best of intentions at the onset of the work. As a result of this low likelihood, construction contracts typically include provisions regarding the administration of disputes, so projects do not grind to a halt when a dispute arises. Having an appreciation of the fact that the design and construction process is an imperfect science due to the sheer number of moving parts and the atypical nature of each project is important for each party to consider. Furthermore, when parties to a construction project do understand and follow the contract terms related to disputes, the overall performance of the project generally improves. The aim of this book is to provide a tool for all parties to improve the dispute administration process to improve the overall performance of projects.

This book is focused on contract claims. A contract claim is just what it sounds like—it's based on a breach of a contract provision of an agreement between the parties related to a construction project. The party asserting the claim is the "claimant" and the party responding to the claim is the "respondent." Contract claims are easy to define if the terms of the contract are clear. If contract terms are ambiguous and require interpretation, that typically requires legal opinions. However, if the four corners of the contract clearly define the duties of the parties to the contract, legal interpretation is not necessary as the contract will speak for itself. If the dispute cannot be resolved at the project level and binding dispute resolution is required, the parties will require attorneys to administrate the arbitration or litigation process, as well as experts to evaluate and opine on dispute issues.

I've developed a nine-step approach to contract claims that will help claimants prepare affirmative claims and will assist respondents to evaluate the merits of affirmative claims. This process is summarized in Figure 1.1.

## I.   Step 1: Review the Dispute Resolution Procedure

Because issues and disputes on construction projects are inevitable, parties to a construction project must understand how these events are administered. All too often, parties to a contract know very little about the dispute resolution terms they are bound to, and this can cause issues and disputes to snowball, which can frustrate the administration of the work.

## II.   Step 2: Define the Type of Dispute

Contracts often have specific procedures and remedies for specific types of disputes, so it is critical to understand the type of dispute at bar. For instance, for contractors, standard contract forms deal differently with owner design issues, owner administrative issues, owner performance issues, third-party issues, and change order negotiation issues. Thus, it is critical to define the nature of the claim in order to understand how the contract deals with specific types of claims.

**DISPUTE RESOLUTION FLOWCHART**

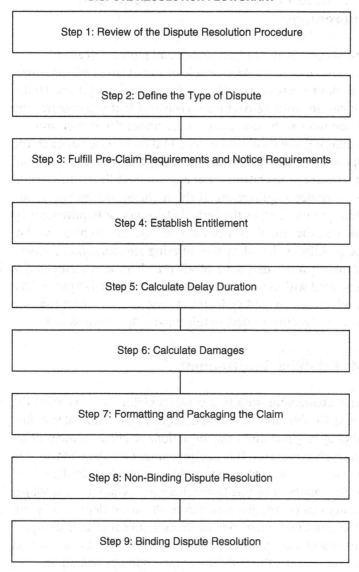

Step 1: Review of the Dispute Resolution Procedure

Step 2: Define the Type of Dispute

Step 3: Fulfill Pre-Claim Requirements and Notice Requirements

Step 4: Establish Entitlement

Step 5: Calculate Delay Duration

Step 6: Calculate Damages

Step 7: Formatting and Packaging the Claim

Step 8: Non-Binding Dispute Resolution

Step 9: Binding Dispute Resolution

**Figure 1.1**  Dispute resolution flowchart.

## III. Step 3: Fulfill Pre-Claim Requirements and Notice Requirements

Once the type of dispute is identified, the claimant must fulfill all pre-claim and claim notice requirements. An example of a pre-claim notice requirement relates to design issues involving differing site conditions. Under an AIA (American Institute of Architects) contract that incorporates the A201 general conditions, the contractor must notice the owner and the architect of a differing site condition within 14 days of observance so the architect can promptly investigate and either deny the claim or propose an equitable adjustment to the contract sum and contract time—this would be the pre-claim notice requirement. If the architect denies the claim, then the contractor must trigger the formal claim notice requirement by noticing the owner, the initial decision maker, and the architect within 21 days of the architect's denial of the differing site condition request. Hence, the claimant must be aware of all pre-claim and claim notice requirements defined within the contract. If the exact notice requirements are not followed, claimants and claimant attorneys often assert that the respondent had constructive notice, so this topic is discussed as well.

## IV. Step 4: Establish Entitlement

This is the most challenging step to a contract claim and it is often the one most ignored. Establishing entitlement requires the claimant to define what is required by contract and how the actions or circumstances differ from the contract requirements. It is not uncommon for claimants to skip or provide a cursory entitlement review and then provide a detailed damages analysis. This methodology is flawed because a claimant cannot get to delay or damages unless the claimant can establish entitlement. Regardless whether the claimant is an owner, contractor, subcontractor, designer, or vendor, entitlement must be established to prove a claim. Depending on the type of issue or dispute, the entitlement methodology will differ.

## V. Step 5: Calculate Delay

Once entitlement is established, the claimant can work on the delay analysis, if applicable. Contract claims often have a delay component, which results in delay damages. In order to establish that an issue or dispute had a critical impact on the claimant's work, the claimant should use one or more of the four accepted forensic scheduling techniques:

1) time impact analysis;
2) windows analysis;
3) collapsed as-built analysis;
4) as-planned vs. as-built analysis.

The selection of the forensic technique depends on the information available to the claimant and whether the claimant seeks to prove an excusable and compensable delay or an excusable non-compensable delay.

## VI.   Step 6: Calculate Damages

Once entitlement is established and the delay is calculated (if applicable), the claimant can move on to damages. The five main types of damages are:

1) scope change damages;
2) productivity damages;
3) acceleration damages;
4) delay damages;
5) consequential damages.

Depending on what cost information exists, damages for the first four categories may be calculated by one of four methods:

1) actual cost method;
2) agreed upon cost method;
3) estimated cost method;
4) modified total cost method.

Consequential damages relate to costs not associated with on-site project activity, such as home office overhead, loss of bonding capacity, and lost profits. Consequential damages are often mutually waived in contracts and are not subject to markups.

## VII.   Step 7: Formatting and Packaging the Claim

The next step involves the packaging of the claim such that it is presented in an organized and clear fashion and it includes all the necessary backup for the reader to reference. Claims that are properly packaged have a much better chance of being resolved in a timely fashion and before binding dispute resolution is required, which saves time and costs for all parties. Chapter 8 will review how claims can be packaged in order to put the claimant in the best position to resolve the dispute.

## VIII. Step 8: Non-Binding Dispute Resolution

Standard construction contracts require non-binding dispute resolution such as settlement meetings at the project level, settlement meeting with decision makers, and formal mediation. Chapter 9 discusses best practices with regards to managing these non-binding forums. If the parties are interested in settling the dispute and the right mediator is in place, most disputes are settled.

## IX. Step 9: Binding Dispute Resolution

If the non-binding dispute resolution process fails, the claimant is left with a binding dispute resolution to resolve the dispute, which is in the form of litigation or arbitration, depending on what the subject contract stipulates. Binding dispute resolution is time-consuming and costly, so the claimant should go into this process with eyes wide open. Chapter 10 discusses best practices in terms of litigating or arbitrating a construction dispute.

## X. Other Topics

This book also includes additional chapters to address other important topics related to construction claims, such as termination claims, non-contract claims, and allocation of damages.

### A. Termination Claims

Termination claims are worthy of a separate chapter because termination is the most severe remedy provided to the parties to the contract and standard construction contract forms define when and how one party can terminate the other party to the contract and what damages are recoverable in the event of a proper termination. Thus, termination clauses are often self-contained in terms of entitlement, procedure, and damages. Chapter 11 reviews these standard provisions and provides examples of both proper and improper terminations.

## B. Non-Contract Claims

While this book is focused on contract claims, many of the concepts noted herein also apply to tort claims. Tort claims are non-contract claims that arise when one party has a duty to other foreseeable parties, and when this duty is breached, and it results in damage, monetary damages and/or performance are owed. An example tort claim related to a construction project is when a Homeowners Association (HOA) sues developers, design firms, contractors, subcontractors, and vendors due to design and construction elements that fall below the requisite standard of care, even though no contract exists between these parties and the HOA. State statutes and state case law typically define tort law doctrine and remedies. In order to establish duties in a tort claim, "standard of care" experts are often retained by the parties that opine on which duties are owed and if conduct met or fell below the requisite standard of care. Some of the information in Chapter 12 applies to tort claims, such as damage calculations, and fault allocation is covered in Chapter 13. Other non-contract claims are highlighted as well, including: (1) quantum meruit; (2) unjust enrichment; (3) negligence; (4) breach of warranty; and (5) mechanic's lien claims. It is important for claimants and respondents to be familiar with these claims as they often arise during binding dispute resolution.

## C. Fault Allocation

If a claim involves one or more parties, the claimant or respondent may need to allocate fault amongst the parties to properly apportion claimed damages. For instance, if an owner has an administrative claim regarding lack of clean-up by both the contractor and a separate contractor on site, the owner must properly allocate the damages between the parties. As discussed in detail in Chapter 13, this process involves a five-step approach: (1) defining the issue; (2) duties; (3) patent or latent in nature; (4) was there a cover-up?; and (5) responsibility.

## XI. Summary

The aim of this book is to assist all parties to a construction project with the dispute resolution process. When the parties understand the dispute resolution process and can properly prepare and evaluate claims during the course of the project, there is a higher likelihood for the dispute to be resolved in a timely fashion and well before binding dispute resolution, which is when significant legal and expert fees are incurred by the parties. Moreover, if a dispute ends up being resolved in a binding dispute resolution format, the party that previously prepared or rebutted the claim in a reasonable fashion during the design and construction of the project will likely be placed in a favorable position during the trial or arbitration.

# 2

## Step 1: Review the Dispute Resolution Procedure

The first step in the claims process is for the claimant to begin with the end in mind. While it is preferred for the claimant to negotiate a reasonable settlement with the respondent well before the binding dispute resolution process is initiated, it is important for the claimant to understand the overall pathway for dispute resolution up front, rather than learn about it along the way—which is too often the case and this can be a costly learning curve. For one thing, an understanding of the overall dispute resolution process allows claimants and respondents to make sound business decisions before the resolution process incurs significant legal, expert, and mediator/arbitrator/court fees. In addition, if the claimant deems the respondent to be irrational, they might decide to expedite the process toward binding dispute resolution.

Nearly all contract agreements that relate to a construction project define the dispute resolution procedures. The construction industry uses either standard or proprietary contract forms. Most agreements related to private construction work use standard contract forms, while most public improvement projects use proprietary contract forms. The three most common standard contract forms that can be purchased online are published by: (1) the American Institute of Architects (AIA), which are primarily used for building projects;[1] (2) the Engineers Joint Contract Documents Committee (EJCDC);[2] which are generally used for horizontal infrastructure projects; and (3) ConsensusDocs[3] that are applicable for both horizontal or vertical construction projects.

Most federal, state, and local agencies use proprietary contract forms that are designed to address specific project types and regulatory requirements. For federal agency work, the dispute resolution procedures follow Federal

Acquisition Regulation (FAR) requirements. State and local agencies stipulate dispute resolution requirements based on preference. Certain private owners also use proprietary contracts forms, particularly those that let a large volume of construction work.

Depending on the contract form, the dispute resolution process can be quick or lengthy in duration. If lengthy dispute resolution provisions are required and the claim value is relatively small, the claimant might not want to involve attorneys and expert consultants until the binding dispute resolution stage, particularly if the contract does not include a prevailing party provision that allows for the recovery of fees. If this is the case, the claimant might win the case, but lose the war because fees can wipe out most, if not all, of an award, and then some. Thus, claim administration involves cash flow forecasting and likely outcome predictions in order to properly manage settlement negotiations throughout the dispute resolution process.

Typical dispute resolution steps involve: (1) the claimant issues a notice of claim to the respondent; (2) the respondent renders a decision on the claim; (3) if the claimant finds that the respondent's position is unacceptable, the claimant triggers a non-binding dispute resolution, such as mediation; and (4) if the non-binding dispute resolution does not resolve the claim, the claimant moves forward with a binding dispute resolution. The timing and specifics for each of these steps vary from contract form to contract form, so it is imperative for the claimant to understand the dispute resolution procedures of the subject agreement.

Dispute resolution provisions also vary depending on the parties to the agreement, so owner–contractor agreements have dispute resolution terms that differ from contractor–subcontractor agreements, and both differ from the terms of owner–designer agreements. Note that the project delivery method used on a given project alters the duties and responsibilities of the parties, so it is important to know whether the delivery method is design-bid-build with a stipulated sum, design-bid-build with a guaranteed maximum price, construction manager for fee, construction manager at risk, design-build, engineer-procure-construct, integrated project delivery, etc.

The following sections review the dispute resolution provisions for various types of agreements.

# I.  Standard Contract Forms for Owner–Contractor Agreements

The following is a list of dispute resolution provisions for the three commonly used standard contract forms for owner–contractor agreements. Note that each form allows the parties to edit the contract before execution, so it is important to review the exact dispute resolution provisions within the subject contract.

## A.  AIA A201 General Conditions, Article 15, Claims and Disputes

The AIA was founded in 1857 and it is a professional organization for architects in the United States. The AIA's main design-bid-build contract forms between an owner and a contractor (AIA A101, A102, and A103 forms) all incorporate the AIA A201 General Conditions, which include the dispute resolution provisions for these contracts. The AIA's primary construction manager as constructor (CMc) contract forms (AIA A133 and A134) also incorporate the AIA A201 General Conditions. Article 15, "Claims and Disputes," of the A201 offers the following dispute resolution steps:

- The A201 designates the project architect as the "Initial Decision Maker," unless otherwise noted in the agreement. The contractor or owner shall issue claims to the Initial Decision Maker for an initial decision on claims within **21 days** after occurrence of the event giving rise to the claim or first recognition of the event giving rise to the claim.
- The Initial Decision Maker shall review claims within **10 days** of receipt and take one or more of the following actions: (1) request supporting data; (2) reject the claim in whole or in part; (3) approve the claim; (4) propose a compromise; or (5) advise the parties that it is unable to render a decision.
- If the Initial Decision Maker does not render a decision within **30 days** of claim submission, the claimant may demand mediation and binding dispute resolution without an initial decision.
- If the Initial Decision Maker requests a party to answer questions or furnish additional supporting data, such party shall respond within **10 days**

upon receipt. Upon receipt of the response or supporting data, if any, the Initial Decision Maker will either reject or approve the claim in whole or in part [presumably within **10 days** from receipt—the A201 does not list a specific duration here].

- Either party may file for mediation of an initial decision at any time. Also, within **30 days** of receipt of the initial decision, one party, which is typically the prevailing party of the initial decision, can demand in writing that the other party file for mediation of the initial decision. If such demand is made and the party receiving the demand fails with file for mediation within 30 days thereafter, then the initial decision becomes binding.

- A party can file for binding dispute resolution concurrently with the filing for mediation. Binding resolution shall be stayed pending mediation for a period of **60 days**, unless a longer period is agreed upon by the parties or court order.

- Either party may, within **30 days** from the date that mediation has been concluded without resolution or **60 days** after mediation has been demanded without resolution, demand in writing that the other party file for binding dispute resolution. If such a demand is made and the party receiving the demand fails to file for binding dispute resolution within **60 days** after receipt, then both parties waive their rights to binding dispute resolution and the initial decision becomes binding.

- The claim is resolved via binding dispute resolution.

## B. ConsensusDocs 200, Standard Agreement and General Conditions Between Owner and Constructor

ConsensusDocs was founded in 2007 by a coalition of architecture-engineering-construction (AEC) organizations. ConsensusDocs' owner–contractor contract forms integrate general terms and general conditions into one document. The dispute resolutions provisions for Consensus-Docs' general contracting series are as follows:

- In the event the project personnel from the contractor and owner cannot resolve a dispute, direct discussions should be conducted between the parties' representatives that possess the necessary authority to resolve the matter.

- If the parties' representatives are not able to resolve the matter within **5 business days** of the first discussion, the parties' representatives shall inform senior executives of the parties that resolution could not be reached.

- Upon receipt of such notice, the senior executives of the parties shall meet within **5 business days** to endeavor to resolve the matter. If the dispute remains unresolved after **15 business days** from the date of the date of the first discussion, the parties shall submit the matter to dispute mitigation, if applicable, and dispute resolution procedures.
- The optional non-binding dispute mitigation procedures can either be a project neutral or a dispute review board. The project neutral or dispute review board shall issue nonbinding findings within **5 business days** of a referral. If the matter remains unresolved after the findings, the parties shall submit the matter to binding dispute resolution.
- If direct discussions do not result in resolution and no dispute mitigation procedure is selected, the parties shall endeavor to resolve the matter by mediation. Mediation shall be convened within **30 business days** of the matter first being discussed and shall conclude within **45 business days** of the matter first being discussed. Either party may terminate the mediation at any time after the first session by written notice to the non-terminating party and mediator.
- The matter is resolved via binding dispute resolution.

## C.  C-700, Standard General Conditions of the Construction Contract (2018 Version)

EJCDC is a joint venture of three major organizations of professional engineers: (1) the American Council of Engineering Companies; (2) the National Society of Professional Engineers; and (3) the American Society of Civil Engineers. Most EJCDC stipulated sum and cost-plus-fee contract forms incorporate EJCDC's C-700 general conditions, which set forth dispute resolution provisions within Article 10, "Changes in the Work; Claims," and Article 16, "Dispute Resolution." The process is summarized as follows:

- If the contractor or owner has a dispute that involves time and/or money, the claimant shall issue written notice to the engineer within **30 days** after the start of the event giving rise to the claim.
- The claimant shall provide the engineer with supporting cost and/or time data within **60 days** after the start of the claim event.

- The respondent shall submit a response to the engineer within **30 days** after receipt of the claimant's last submittal.
- Within **30 days** of the respondent's last submittal, the engineer will either approve the claim, deny the claim, or advise the parties it cannot resolve the claim.
- Either party can invoke the dispute resolution procedure within **30 days** of the engineer's decision. Failure to do so within this 30-day period renders the engineer's decision final and binding.
- The owner or contractor can request mediation of a claim issue to the engineer for a decision before it becomes final and binding. The parties shall conduct mediation within **60 days** of filing.
- Within **30 days** of an unsuccessful mediation, the contractor can trigger binding dispute resolution.
- The claim is resolved via binding dispute resolution.

## D.  Proprietary Contract Dispute Resolution Provisions for Owner–Contractor Agreements

Proprietary contract forms are used by thousands of public and private owners so it is not possible to review the dispute resolution provisions for each proprietary contract form; however, it is possible to review the dispute resolution procedures that apply to federal construction contracts as well as a typical state contract, in order for the reader to get a feel for the typical requirements.

### 1.  Federal Projects Dispute Resolution Provisions for Owner–Contractor Disputes

The US Federal Government procures and administers construction contracts according to the Federal Acquisition Regulation (FAR). FAR Regulation Part 33, Subpart 33.2, "Disputes and Appeals," outlines the dispute resolution procedures for federal construction contracts. Section 33.202, "Disputes" notes that 41 U.S.C. chapter 71, "Disputes," establishes procedures and requirements for asserting and resolving claims.

- A contractor shall submit claims against the Federal Government to the contracting officer for a decision. For claims of more than $100,000, the contractor shall certify that the claim is made in good faith, the supporting data are accurate and complete to the best of the contractor's knowledge and belief, the amount requested accurately reflects the contract adjustment for which the contractor believes the Federal Government is liable, and the certifier is authorized to certify the claim on behalf of the contractor.

- The contracting officer shall issue a decision in writing to the contractor. For claims less than $100,000, the contracting officer shall issue a decision within **60 days**. For claims over $100,000, the contracting officer shall issue a decision within **60 days or notify the contractor when the decision will be issued**.
- A contractor can appeal a contracting officer's decision to an Agency Board[4] within **90 days** of receipt of the decision or bring an action on the claim in the US Court of Federal Claims within **12 months** of receipt of the decision.
- A contractor may appeal Agency Board or US Court of Federal Claims decisions to the US Court of Appeals for the Federal Circuit within **120 days** from the date the contractor receives a copy of the decision.

### 2.   Example State Agency Contract Dispute Resolution Provisions for Owner–Contractor Disputes

Dispute resolution provisions for state construction contracts vary by state and by state agency. The following example is the dispute resolution provisions that the Colorado Department of Transportation prescribes in its standard construction contracts.

*Colorado Department of Transportation -- 2019 Standard Specifications for Road and Bridge Construction*   CDOT's "Standard Specifications for Road and Bridge Construction" include exhaustive provisions for non-binding dispute resolution. This non-binding process can take more than a year before a contractor can trigger binding dispute resolution. The process is outlined as follows:

- Contractor to bring dispute issue to CDOT's Project Engineer's attention within **20 days** of Contractor being aware of the issue.
- If this discussion results in an impasse, Contractor issues a notice of dispute to CDOT's Project Engineer within **15 days** of the impasse.
- Contractor provides Request for Equitable Adjustment (REA) to the CDOT's Project Engineer within **15 days** of the notice of dispute.
- Contractor and the CDOT's Project Engineer discuss merits of dispute within **7 days** of the REA submission.

- CDOT's Project Engineer reviews the REA within **7 days** of the discussion.
- Contractor rejects the CDOT's Project Engineer's denial and notices the CDOT's Resident Engineer within **7 days** of the Project Engineer's rejection.
- Contractor's project and executive team meet with the CDOT's Project Engineer and the Resident Engineer to discuss within **30 days** of notice.
- If impasse, the CDOT's Project Engineer initiates Dispute Resolution Board (DRB) process within **30 or 45 days** of this group meeting, depending on type of DRB.
- Contractor to sign the DRB agreement within **20 days**.
- Contractor and the CDOT issue pre-hearing submittals to DRB within **15 days** of the signing of the agreement.
- DRB, the Contractor, and the CDOT to conduct hearing within **30 days** of the submittal issuance.
- DRB renders a "recommendation" within **10 days** of the hearing.
- Parties can issue request for clarification/reconsideration of recommendation within **14 days** of the DRB's recommendation.
- Either party can reject the DRB's recommendation within **30 days**.
- Claimant to issue notice of intent to file "claim" to the Region Transportation Director within **60 days** of the DRB's recommendation.
- Contractor submits claim package to the Region Transportation Director within **60 days** of notice.
- Region Transportation Director renders a decision within **30 days** of receipt of the claims package.
- Contractor can reject the Region Transportation Director's decision and issue an appeal to the CDOT's Chief Engineer within **60 days** of the Region Transportation Director's decision.
- CDOT's Chief Engineer shall render its decision within **45 days** of the appeal.
- Contractor can reject the CDOT's Chief Engineer's decision and trigger binding dispute resolution within **180 days** of the Chief Engineer's decision.
- The claim is resolved via binding dispute resolution.

## II.  Standard Contract Forms for Contractor–Subcontractor Agreements

The following is a list of dispute resolution provisions for the three most commonly used standard contract forms for contractor–subcontractor agreements. Note that each form allows the parties to edit the contract before execution, so it is important to review the exact dispute resolution provisions within the subject contract.

### A.   AIA A401, Article 6, Claims and Disputes

The AIA Document A401 – 2017, "Standard Form of Agreement Between Contractor and Subcontractor," is the AIA's most commonly used form of agreement between a contractor and a subcontractor. Article 6, "Claims and Disputes," of the A401 offers the following dispute resolution procedures:

- Claims shall be made promptly and per the terms of the subcontract agreement.
- Claims between the parties are subject to mediation as a condition precedent to binding dispute resolution. The request for mediation shall be in writing, delivered to the other party, and filed with the person or entity administering the mediation. The parties shall share the mediator's fees equally.
- A party can file for binding dispute resolution concurrently with the filing for mediation. Binding resolution shall be stayed pending mediation for a period of **60 days**, unless a longer period is agreed upon by the parties or court order.
- The method of binding dispute resolution shall be arbitration, litigation, or otherwise selected within the A401 agreement. If no method is selected, claims shall be resolved by litigation in a court of competent jurisdiction.

### B.   ConsensusDocs 750, Standard Agreement Between Constructor and Subcontractor

The ConsensusDocs 750, "Standard Agreement Between Constructor and Subcontractor," is ConsensusDocs' most commonly used form of agreement between a constructor and a subcontractor. Article 11, "Dispute Mitigation and Resolution," of the 750 offers the following dispute resolution procedures:

- In the event the project personnel from the constructor and subcontractor cannot resolve a dispute, direct discussions should be conducted between the parties' representatives that possess the necessary authority to resolve the matter.
- If the parties' representatives are not able to resolve the matter within **5 business days** of the first discussion, the parties' representatives shall inform senior executives of the parties that resolution could not be reached.
- Upon receipt of such notice, the senior executives of the parties shall meet within **5 business days** to endeavor to resolve the matter. If the dispute remains unresolved after **15 business days** from the date of the date of the first discussion, the parties shall submit the matter to mediation.
- Mediation shall be convened within **30 business days** of the matter first being discussed and shall conclude within **45 business days** of the matter first being discussed. Either party may terminate the mediation at any time after the first session by written notice to the non-terminating party and mediator.
- The matter is resolved via binding dispute resolution via arbitration or litigation, whichever is selected. Note the 750 agreement has a prevailing party provision in terms of attorney fees.

## C.  EJCDC E-523, Construction Subcontract Agreement (2018 Version)

The EJCDC's E-523 contract form, "Construction Subcontract," is the EJCDC's most commonly used form of agreement between a contractor and subcontractor. Article 12, "Claims and Dispute Resolution," of the E-523 offers the following dispute resolution procedures:

- Subcontractor shall issue a notice of claim to the contractor within **30 days** from the event giving rise to the claim, and for claims related to the prime contract, within **5 days**.
- Either party may request mediation of any unresolved dispute. The mediation process must be concluded within **60 days** of filing of the request.

- If the dispute is not resolved by mediation, each party is barred from further action to assert its claim after **30 days** after termination of the mediation, unless, within this period of time, a party elects in writing to invoke dispute resolution or gives the other party written notice of its intent to submit the claim to a court of competent jurisdiction.

## III.  Standard Contract Forms for Owner–Designer Agreements

The following is a list of dispute resolution provisions for the three most commonly used standard contract forms for owner–designer agreements. Note that each form allows the parties to edit the contract before execution, so it is important to review the exact dispute resolution provisions within the subject contract.

### A.  AIA B101, Article 8, Claims and Disputes

The AIA Document B101 – 2017, "Standard Form of Agreement Between Owner and Architect," is the AIA's most commonly used form of agreement between an owner and an architect. Article 8, "Claims and Disputes," of the B101 offers the following dispute resolution procedures:

- The architect or owner shall commence all claims and causes of action against the other within the period specified by applicable law, but not longer than **10 years** after the date of substantial completion of the applicable project.
- Claims between the architect and the owner are subject to mediation as a condition precedent to binding dispute resolution. The request for mediation shall be in writing, delivered to the other party, and filed with the person or entity administering the mediation. The parties shall share the mediator's fees equally.
- A party can file for binding dispute resolution concurrently with the filing for mediation. Binding resolution shall be stayed pending mediation for a period of **60 days** unless a longer period is agreed upon by the parties or court order.
- The method of binding dispute resolution shall be arbitration, litigation, or otherwise selected within the B101 agreement.

### B. ConsensusDocs 240, Standard Agreement Between Owner and Design Professional

The ConsensusDocs 240, "Standard Agreement Between Owner and Design Professional," is ConsensusDocs' most commonly used form of agreement between an owner and a designer. Article 9, "Dispute Mitigation and Resolution," of the 240 offers the following dispute resolution procedures:

- In the event that project personnel from the design professional and owner cannot resolve a dispute, direct discussions should be conducted between the parties' representatives that possess the necessary authority to resolve the matter.
- If the parties' representatives are not able to resolve the matter within **5 business days** of the first discussion, the parties' representatives shall inform senior executives of the parties that resolution could not be reached.
- Upon receipt of such notice, the senior executives of the parties shall meet within **5 business days** to endeavor to resolve the matter. If the dispute remains unresolved after **15 business days** from the date of the date of the first discussion, the parties shall submit the matter to dispute mitigation, if applicable, and dispute resolution procedures.
- The optional non-binding dispute mitigation procedures can either be a project neutral or a dispute review board. The project neutral or dispute review board shall issue nonbinding findings within **5 business days** of a referral. If the matter remains unresolved after the findings, the parties shall submit the matter to binding dispute resolution.
- If direct discussions do not result in resolution and no dispute mitigation procedure is selected, the parties shall endeavor to resolve the matter by mediation. Mediation shall be convened within **30 business days** of the matter first being discussed and shall conclude within **45 business days** of the matter first being discussed. Either party may terminate the mediation at any time after the first session by written notice to the non-terminating party and mediator.
- The matter is resolved via binding dispute resolution (arbitration or litigation).

## C. EJCDC E-500, Agreement Between Owner and Engineer for Professional Services (2020 Version)

The EJCDC's E-500 contract form, "Agreement Between Owner and Engineer for Professional Services," is the EJCDC's most commonly used form of agreement between an owner and an engineer. Section 6.07, "Dispute Resolution," of the E-500 offers the following dispute resolution procedures:

- Owner or engineer shall issue a notice of claim to the other party. Within **30 days** from notice, the parties shall attempt to negotiate all disputes in good faith.
- After 30 days, the owner and the engineer shall submit any unsettled disputes to mediation. The mediation process must be completed within **120 days**.
- If the dispute remains unresolved after mediation, either party may invoke binding dispute resolution. If Exhibit H is included within the E-500 agreement, arbitration is stipulated and must be filed before the dispute would be barred by the applicable statute of limitations. If the E-500 agreement does not incorporate Exhibit H, venue shall be state court having jurisdiction at the location of the project (or federal court in the district in which the project is located, if appropriate).

# IV. Standard Purchase Order Forms for Purchaser–Vendor Agreements

The following is a list of dispute resolution provisions for standard purchase order forms for Purchaser–Vendor Agreements published by the AIA and ConsensusDocs—the EJCDC does not have a standard purchase order template. Note that each form allows the parties to edit the purchase order before execution, so it is important to review the exact dispute resolution provisions within the subject purchase order.

### A. AIA A152 and A152 Exhibit A, Article 8, Claims and Disputes

The AIA Document A152 – 2019, "Purchase Order," is AIA's most commonly used purchase order form between a purchaser and a vendor. This one-page form lists AIA Document A152-2019, Exhibit A – Terms and Conditions, as a contract document. Section 1.2 of Exhibit A, "Governing Law, including the Uniform Commercial Code," notes the choice of law as the project jurisdiction's Uniform Commercial Code (UCC) as adopted, and it indicates that disputes are to be resolved in a court of competent jurisdiction unless the parties agree otherwise—no further procedure is provided beyond this statement. Keep in mind that nearly all US jurisdictions have adopted the UCC in some form.

### B. ConsensusDocs 702 and 703, Purchase Orders

ConsensusDocs has two purchase order forms, one for commodity goods (ConsensusDocs 702, which includes a 702.1 attachment) and the other for non-commodity goods (ConsensusDocs 703). Both forms list the parties as the "Buyer" and "Seller" and unlike the AIA form, the ConsensusDocs' forms provide detailed dispute resolution procedures. The 702 lists the following dispute resolution procedures:

- In the event that a dispute arises, direct discussions should be conducted between the parties' representatives that possess the necessary authority to resolve the matter.
- If the parties' representatives are not able to resolve the matter within **5 business days** of the first discussion, the parties' representatives shall submit the matter to binding dispute resolution, which can be litigation or arbitration depending on what the parties selected on the purchase order form.

The 703 form adds a mediation requirement before binding dispute resolution:

- In the event that a dispute arises, direct discussions should be conducted between the parties' representatives that possess the necessary authority to resolve the matter within **5 business days** of the first discussion.
- If the parties fail to resolve the matter through direct discussions, the dispute shall be submitted to mediation pursuant to AAA rules. The

parties shall select a mediator within **15 business days** of the request for mediation.

- If the matter is unresolved after mediation, the parties' representatives shall submit the matter to binding dispute resolution, which can be litigation or arbitration depending on what the parties selected on the purchase order form.

## V.  Summary

The first step in perfecting an affirmative claim is to review the dispute resolution requirements of the subject contract. While standard contract forms are editable, the procedures for dispute resolution are generally the same from contract to contract. To the contrary, the dispute resolution procedures for proprietary contract forms vary. Moreover, dispute resolution terms vary based on the type of contract form. To wit, owner–contractor forms terms vary from contractor–subcontractor form terms, and these forms vary from owner–designer contract forms and purchaser–vendor forms. Thus, it is critical for claimants to understand the steps that are necessary to notice claims, get a project-level decision, run through non-binding dispute resolution requirements, and then trigger a binding dispute resolution process. As time goes on, the cost to administer claims goes up, so presenting proper and clearly supported claims early in the process generally saves both parties time and money.

## Notes

1  https://www.aiacontracts.org/
2  https://www.ejcdc.org/online-store/
3  https://www.consensusdocs.org/
4  Agency Boards include: (1) the Armed Services Board for contracts with military agencies; (2) the Post Service Board for contracts with the US Postal Service or Postal Regulatory Commission; and (3) the Civilian Board for all other federal agencies unless otherwise specified.

# 3

## Step 2: Define the Type of Dispute

Contract claims generally fall into five different categories. Standard contract forms have varying notice requirements and allowable remedies for each of these categories. Thus, it is important for the claimant to understand which category the claim falls into so the claim can be administered accordingly. The five claim categories are:

1) **Design Issues:** When a party is impacted by a design issue on the project and the other party to the contract is responsible for the subject design.
2) **Administration Issues:** When a party is impacted by the maladministration of the other party to the contract.
3) **Performance Issues:** When a party is impacted by the poor performance of the other party of the contract.
4) **Third-Party Issues:** When a party is impacted by a third party and can only recover if the other party to the contract is able to pass the claim through to the third party and recover from the third party. Or, if one or both parties to the contract are impacted by a force majeure event.
5) **Change Order Negotiation Issues:** When both parties to a contract agree that a change should be made to the contract sum and/or contract time but the parties fail to agree to the amount of the change.

## A.  Design Issues

Many parties can be impacted by design issues. For instance, a contractor may be impacted by a design error and if the owner pays the contractor additional funds and grants additional time, the owner is then impacted and might seek recovery from the designer, and then the designer might then seek recovery from a responsible subconsultant. Or, an owner might delegate certain portions of the design to the contractor and if the design generated by the contractor is deficient, the owner might be the impacted party and seek recovery from the contractor via a backcharge. The contractor might then seek recovery from its retained design firm.

Another scenario might be that an owner is impacted by its design builder's deficient design program and this might lead to an owner backcharge to the design builder. The design builder might then assert a claim against the designer. Thus, the scenarios are numerous, but the key point here is that a design issue exists when one party relies upon the design of the other party and when the design is deficient or revised without consideration, the impacted party seeks recovery from the other party. The most common design issues are as follows:

- **Differing Site Conditions:** When one party relies upon the site conditions represented by the other party and the represented conditions turn out to be different, which impacts costs and/or schedule. Or, when the contract documents are silent regarding the site conditions and the site conditions differ from what is generally found at the project location and this impacts a performing party's costs and/or schedule.
- **Added Scope:** When one party relies upon the design provided by the other party and the design later expands and the expanded design impacts the costs and/or schedule of the party that relied upon the original design.
- **Changed Scope:** When one party relies upon the design provided by the other party and the design later changes and the changed design impacts the costs and/or schedule of the party that relied upon the original design.
- **Design Errors:** When one party relies upon the design provided by the other party and the party that relies upon the design identifies flaws in the design that will cause cost and/or time impacts to its work.
- **Design Omissions:** When one party relies upon the design provided by the other party and the party that relies upon the design identifies omissions in the design that will cause cost and/or time impacts to its work.
- **Late Design Issuance:** When one party relies upon the design provided by the other party and the party that is responsible for the design is late in issuing its complete design and this causes cost and/or time impacts to the other party's work.

- **Excessive Changes:** When one party relies upon the design provided by the other party and the party that relies upon the design is faced with numerous design changes during construction of the project, which causes a cumulative impact to its work.
- **Cardinal Changes:** When one party relies upon the design provided by the party and the party that is responsible for the design fundamentally alters the scope of the other party's work via a change or changes in the design, which frustrates or makes impossible the purpose of the agreement by significantly increasing the volume of the work, by drastically modifying the type of work, and/or by greatly increasing the original contract costs.

## B. Administration Issues

Standard contract forms define the administrative duties required by each party to the agreement. For instance, on a design-bid-build stipulated sum agreement between a contractor and an owner, each party relies upon the other party to timely and properly administer payment applications, submittals, requests for information (RFIs), inspections, and other administrative duties. When one of the parties fails to administer the contract as required, and this detrimentally impacts the other party's cost and/or time on a project, the impacted party can assert a maladministration claim against the other party. Such claims are often difficult to quantify contemporaneously—as the claimant often has to establish a pattern of maladministration—so impacts are frequently determined retrospectively, after much or all of claimant's work is complete. Typical maladministration claims include:

- **Payment Administration Issues:** When one party maladministers the payment process and this impacts the other party's costs and/or time on the project.
- **Proof of Funding Issues:** When the paying party is required to show the performing party proof of adequate and its failure or delay in doing so impacts the performing party's costs and/or time on the project.
- **RFI Administration Issues:** When one party maladministers the RFI process and this impacts the other party's costs and/or time on the project.

- **Submittal Administration Issues:** When one party maladministers the submittal process and this impacts the other party's costs and/or time on the project.
- **Change Order Administration Issues:** When one party maladministers the change order process and this impacts the other party's costs and/or time on the project.
- **Inspection Issues:** When one party maladministers the inspection process and this impacts the other party's costs and/or time on the project.
- **Subcontractor Issues:** When one party maladministers the process of submitting or approving subcontractors and this impacts the other party's costs and/or time on the project.
- **Project Management Team Issues:** When one party maladministers the process of submitting or approving project management team members and this impacts the other party's costs and/or time on the project.
- **QA/QC Issues:** When one party maladministers the quality assurance/quality control process and this impacts the other party's costs and/or time on the project.
- **Safety Administration Issues:** When one party maladministers the safety process and this impacts the other party's costs and/or time on the project.

Certain maladministration claims are easy to pinpoint, such as delinquent payments or lack of proof of funding. For these issues, contracts typically allow the claimant to halt work if respondent payments or proof of funding are not completed within a certain number of days after the claimant's notice to respondent. For other maladministration issues, it can be challenging to point to a single event as resultant impacts are often cumulative in nature. Cumulative impacts are analogous to death by a thousand cuts. Cumulative impacts might involve the excessive turnaround time of requests for information (RFIs), submittals, or change orders. Another example of a cumulative impact claim would be inspection delays or over-inspection issues.

## C.  Performance Issues

Performance issues involve claims made by one party to the contract that assert that the performance of the other party (or third parties that the other party is responsible for, such as subcontractors and vendors) is noncompliant with the contract requirements. However, there are instances when the paying party to a contract takes on performance obligations, such as furnishing materials to the performing party or completing work that is a predecessor to the performing party's work. When a party fails to perform its physical work in accordance with the contract terms, and this failure affects the other party's cost and/or time on a project, the affected party can assert an affirmative claim to seek recovery per the contract terms. Typical performance issues involve:

- **Quality Issues:** When the physical work of one party is noncompliant with the contract requirements and it impacts the other party's costs and/or time on the project.
- **Schedule Issues:** When the work of one party is not performed per the contract schedule and this impacts the other party's costs and/or time on the project.
- **Separate Contractor Issues:** When the physical work of one of the party's subcontractors, vendors, or separate contractors is noncompliant with the contract requirements and the contract makes this party responsible for its entities, and one of these entities impacts the other party's costs and/or time on the project.

## D.   Third-Party Issues

Third-party issues typically fall into one of two groups. First, when a party to an agreement, such as a subcontractor in a contractor–subcontractor agreement, is impacted by an owner–designer issue, most contracts allow the subcontractor to submit a claim to the contractor and then the contractor is obligated to pass the claim through to the owner for attempted recovery. Standard contract agreements have strict procedural requirements for this type of claim. Second, when one or both parties to an agreement are impacted for reasons that are out of the control of both parties to the contract, such as a hurricane, certain contract forms allow additional time as the exclusive remedy for the claimant, while others keep the door open for the recovery of both cost and time impacts related to the impact. For this second group of impacts, standard contract forms also have strict procedural requirements such as the source for weather data, etc. Examples of third-party issues include:

- **Pass Through Issues:** When the agreement between two parties contemplates pass through claims and when the performing party that is able to submit a pass through claim has cost and/or time impacts due to a pass through issue, such as a design error.

- **Abnormal Weather Issues:** When the agreement between the parties contemplates average weather conditions and when the performing party has cost and/or time impacts due to abnormal weather conditions.
- **Force Majeure Issues:** When an event that is beyond the control of the parties to a contract either precludes or postpones performance under the contract and this impacts one or both of the parties' costs and/or time on the project.

## E.  Change Order Negotiation Issues

When both parties to an agreement relating to a construction project agree that the claimant is entitled to a change order or backcharge under the agreement but the parties cannot agree to the cost and/or time impacts sought by the claimant, dispute resolution provisions are often triggered by the claimant to resolve time and/or cost impacts.

## Summary

Most claimants wait until the end of the project to move forward with a dispute resolution even though most contract forms consider that such issues will be resolved during the project. Claimants generally take this position to help keep the peace while work is underway; however, this strategy simply kicks the proverbial can down the road and often leads to a nuclear conclusion of the project. Moreover, many standard contract forms allow the paying party to direct the performing party to complete change order work while the cost and time impacts are resolved—this type of direction can place cash flow strains on the performing party and again cause large disputes at the tail end of projects.

**4**

## Step 3: Fulfill Pre-Claim Requirements and Notice Requirements

Construction contracts often prescribe both pre-claim requirements and formal claim notice requirements. In addition, most contracts require that notice be issued in writing and transmitted via an accepted method, such as certified mail or registered mail, archaic as that may seem. Contractor pre-claim notice requirements allow the respondent the opportunity to render a decision on the issue before it is elevated to a formal "claim" status. Formal claim notice generally comes after the respondent has rejected all or a portion of claimant's change request.

It is common for claimants to ignore or otherwise fail to adhere to pre-claim notice requirements and/or formal claim notice requirements. Because lack of notice defenses can cause draconian results for claimants, state and federal courts have ample case law on this subject. States vary on their treatment of respondent's "failure to provide proper notice" defense. A minority of states interpret notice provisions strictly and often time bar delinquent claims, regardless of merit. Most states and federal courts take a more liberal "fairness" approach and often accept claimant defenses that can demonstrate that the respondent:

1) had constructive notice;
2) was not prejudiced due to the lack of notice; or
3) waived notice requirements based on prior conduct.

Claimants that fail to adhere to strict notice provisions and whose claims are rejected by respondents based on a lack of notice should contact a qualified construction attorney to discuss available defenses.

Constructive notice means that the respondent knew or should have known about the claim event. This may include verbal notice, discussions at a meeting(s), or superior knowledge. Second, claimants can assert a lack of prejudice defense if the respondent was not prejudiced by lack of timely notice. In other words, the respondent would not have acted differently even if the claimant had issued proper notice, e.g., the claimed work had to get done for the project to move forward and there was no alternative. Third, a waiver of notice defense may occur when the respondent did not previously enforce strict notice requirements as a condition precedent for the negotiation of claims.

## I.  Pre-Claim Requirements

Standard contract agreements contemplate various types of claims that frequently arise on construction projects and for certain categories of claims these contract forms include pre-claim provisions that allow respondents an opportunity to review the impact and, if so determined, issue a change in the claimant's contract sum and/or contract time before the claimant initiates dispute resolution. Thus, it is important for the claimant to define the nature of a potential claim so it can evaluate whether any pre-claim steps are necessary before it invokes the dispute resolution process. Below is a sample of impacts that often require a pre-claim notice. Each of the examples below is based on a contractor being the claimant and an owner being the respondent and these two parties have entered into a standard contract form and the owner is responsible for the design on the project.

### A.  Design Issues

#### 1.  Differing Site Conditions

A differing site condition occurs when the contractor discovers subsurface or concealed conditions that differ materially from those indicated in the contract documents (Type 1 Differing Site Condition) or materially different from conditions ordinarily encountered in work in the location of the project (Type 2 Differing Site Condition), and this differing condition will cause an increase in the contractor's cost and/or time to perform the work. Standard contracts typically require the contractor to allow the owner the opportunity to decide on whether to authorize a change order to cover resultant impacts. If the contractor does not agree with the owner, it can initiate dispute resolution. The examples below review specific pre-claim requirements.

- **AIA A201 – Section 3.7.4:** The contractor shall provide notice to the owner/architect before conditions are disturbed and no later than 14 days after the first observance of the conditions. The architect shall promptly review the condition and if it agrees with the contractor it will recommend an equitable adjustment in the contract sum or contract time, or both. If the contractor or the owner disputes the architect's position, that party may initiate the Article 15 claims process.

- **ConsensusDocs 200 – Section 3.16.2 and Section 8.2.1:** Per Section 3.16.2, if the contractor identifies a condition at the site that is materially different than indicated in the contract documents or materially different from conditions ordinarily encountered in work in the contract documents, the contractor shall stop the affected work and give prompt written notice of the condition to the owner and the design professional. The owner shall investigate the matter and issue an interim directive that specifies the extent to which the owner agrees that a concealed or unknown condition exists and shall direct the contractor on how to proceed. Per Section 8.2.1, the owner's interim directive may direct the contractor to perform the work before agreeing on an adjustment in the contract price or contract time, or direct the contractor to perform work that the owner believes is not a change. If the owner's interim directive notes the owner's disagreement that a change exists, the contractor can move on to the dispute mitigation and resolution procedures found in Article 12. The contractor can also move on to Article 12 if it disagrees with the owner on the proposed adjustment in the contract price or contract time.

- **EJCDC C-700 – Section 5.04:** If the contractor identifies a subsurface or physical condition that differs materially from that shown or indicated in the contract documents or differs materially from conditions ordinarily encountered and generally recognized as inherent in work of the character provided for in the contract documents, the contractor shall promptly halt the affected work and notify the owner and the engineer in writing about such condition. The engineer is to then promptly review the situation and make a recommendation to the owner. The owner is to then issue a written statement to the contractor that indicates whether any adjustment will be made to the contract. Upon receipt of the owner's statement, the contractor may submit a change proposal to the owner for any adjustment in the contract price or contract time no later than

30 days after the owner's issuance of the statement. If the contractor disagrees with the owner's statement, it can initiate the dispute resolution process per Article 12.

### 2. Other Design Issues

Other design issues include scenarios where the contractor discovers design additions, changes, errors, or omissions in the contract documents that will cause an increase in the cost and/or time to perform the work. Certain contracts note that the contractor should advise the owner of the issue so the owner can render a position on the issue before the contractor initiates dispute resolution. The examples below highlight specific pre-claim requirements.

- **AIA A201 – Section 3.2.2-4:** Before starting each portion of the work, the contractor shall review, in the contractor's capacity as a contractor, the various contract documents and field conditions and report to the architect any errors, omissions, or inconsistencies in the contract documents via a request for information. If upon review of the architect's response to the contractor's request for information, the contractor believes it will cause an increase in the cost and/or time to perform the work, then the contractor shall file a claim per Article 15.
- **AIA A201 – Section 4.2.11-14:** The architect shall promptly respond to requests for information about the contract documents. Thus, before a design claim is issued, it is prudent for the contractor to get the architect's response to a request for information.
- **AIA A201 – Section 8.3.1:** If the contractor is delayed by: actions of owner/architect/separate contractor; changes ordered in the work; force majeure events; abnormal weather; and/or other causes beyond the contractor's control, then the contract time shall be extended as determined by the architect. Section 8.3.3 notes that Section 8.3 does not preclude the contractor's recovery of delay damages from the owner under other provisions of the contract (namely, Article 15). Thus, the contractor should give the architect the ability to render a decision before a claim is initiated.
- **ConsensusDocs 200 – Section 3.3:** Before commencing the work, the contractor shall review the contract document and report any visible conflicts to the owner. Upon receipt, the owner shall promptly inform the contractor on what action, if any, the contractor shall take.

- **EJCDC C-700 – Section 3.03.A.1-2:** If, before or during the performance of work, the contractor discovers a design issue, the contractor shall promptly report it to the engineer in writing. The contractor is not to proceed with the affected work until the matter is resolved.
- **EJCDC C-700 – Section 3.04.A-B:** During the performance of work, the contractor shall issue questions in writing (requests for information) regarding the design to the engineer as soon as possible after such matters arise. The engineer to respond to the contractor's questions with reasonable promptness. Based on the engineer's response, the contractor can submit a change proposal to the owner.
- **EJCDC C-700 – Section 11.09.B:** The contractor shall submit change proposals to the engineer within 30 days after the start of the event giving rise thereto. Note that the change proposal shall comply with Section 4.05.D–E. The engineer is to render an initial decision on the change proposal within 30 days of proper submission. If the contractor does not agree with the engineer's initial decision, it can move to the dispute resolution procedures in Article 12.

## B.   Administration Issues

### 1.   Submittal Issues

Certain contract forms require the contractor to issue a submittal schedule for approval before the contractor can assert delay claims regarding the owner's delinquent review of submittals. This typically takes place shortly after the execution of the contract when the contractor is assembling its schedule for the project. The underlined text below represents specific pre-claim requirements.

- **AIA A201 – Section 3.10.2:** <u>Promptly after being awarded the contract, the contractor to submit a submittal schedule to the architect for the architect's approval.</u> If the contractor fails to submit a submittal schedule or fails to provide submittals in accordance with the approved submittal schedule, the contractor shall not be entitled to any increase in the contract sum or contract price based on the time required for review of submittals.
- **ConsensusDocs 200:** ConsensusDocs 200 does not have a requirement for the contractor to issue a separate submittal schedule. However,

Section 3.14.1 notes that the contractor shall deliver submittals to the owner and the design professional per the project schedule. Section 3.14.2 indicates that the owner is responsible for the review of submittals within reasonable promptness to avoid causing delays. Thus, the contractor should incorporate submittal submissions and submittal returns into its project schedule.

- **EJCDC C-700 – Section 2.03.A.2:** Within 10 days after the effective date of the contract, the contractor shall transmit a submittal schedule to the engineer for approval. Section 7.16.C requires the engineer to review the submittals per the approved schedule of submittals. Thus, the contractor's claims related to delinquent submittal review should be based upon the approved schedule of submittals.

## 2. Other Administrative Issues

Most standard construction contracts require an owner decision on administrative issues before an issue should be moved to the claims process. The underlined text below represents the specific pre-claim requirements.

- **AIA A201 – Section 8.3.1:** If the contractor is delayed by actions of owner/architect/separate contractor, etc., then the contract time shall be extended as determined by the architect. Section 8.3.3 notes that Section 8.3 does not preclude the contractor's recovery of delay damages from the owner under other provisions of the contract, such as Article 15. Thus, the contractor should give the architect the ability to render a decision before a claim is initiated.

- **ConsensusDocs 200:** For all other administrative issues, this document refers to Section 8.4, which details formal claim notice requirements. Accordingly, no other pre-claim notice requirements are prescribed.

- **EJCDC C-700 – Section 11.09.B:** The contractor shall submit change proposals to the engineer within 30 days after the start of the event giving rise thereto. Note that the change proposal shall comply with Section 4.05.D–E. The engineer is to render an initial decision on the change proposal within 30 days of proper submission. If the contractor does not agree with the engineer's initial decision, it can move to the dispute resolution procedures in Article 12.

## C. Performance Issues

### 1. Claim for Negligent Acts or Omissions by the Owner that Cause Injury or Damage to the Contractor

When the owner's negligent acts or omissions cause damage to the contractor's person or property, a minority of standard contracts require the contractor to provide notice to the owner, so the owner is provided an opportunity to investigate and possibly mitigate damage. The examples below outline the specific pre-claim requirements.

- **AIA A201 – Section 10.2.8:** If the contractor suffers injury or damage to person or property because of an act or omission of the owner, the contractor shall give notice to the owner no longer than 21 days after discovery so the owner can investigate the matter.
- **ConsensusDocs 200:** Section 10.1.2 notes that the owner shall indemnify the contractor from all claims from bodily injury and property damage by the negligent or wrongful acts of the owner or design professionals, but this form does not require contractor to provide the owner with a specific notice.
- **EJCDC C-700:** Section 7.16.A–B indicates that the owner shall indemnify the contractor from negligent actors or omissions by the owner and engineer, but this form does not require the contractor to provide the owner with a specific notice.

### 2. Other Owner Performance Issues

Certain standard construction contracts allow the owner to render a decision on performance issues before it moves to dispute resolution. The underlined text below represents specific pre-claim requirements.

- **AIA A201 – Section 8.3.1:** If the contractor is delayed by actions of the owner/architect/separate contractor, etc., then the contract time shall be extended as determined by the architect. Section 8.3.3 notes that Section 8.3 does not preclude the contractor's recovery of delay damages from the owner under other provisions of the contract, such as Article 15. Thus, the contractor should give the architect the ability to render a decision before a claim is initiated.

- **ConsensusDocs 200 – Section 3.6:** The contractor shall examine all items furnished by the owner and shall report any defects in materials or equipment at once to the owner.
- **EJCDC C-700 – Section 8.01.E:** If the contractor's work depends on the work performed by the owner or others, the contractor shall inspect this work and promptly report to the engineer in writing any delays, defects, or deficiencies in such other work.

## D.  Third-Party Issues

Certain standard construction contracts require an owner decision on third-party impacts before such a claim moves to dispute resolution. The underlined text below represents specific pre-claim requirements.

- **AIA A201 – Section 8.3:** If the contractor is delayed by force majeure events, abnormal weather, and/or other causes beyond the contractor's and the owner's control, etc., then the contract time shall be extended as determined by the architect. Thus, the contractor should give the architect the ability to render a decision before a claim is initiated. If the contractor disagrees with the architect's position, the contractor shall issue a claim per Article 15.
- **ConsensusDocs – Section 6.3 and Section 8.4:** Section 6.3 covers, among other things, contractor impacts from third-party issues such as labor disputes, acts of terrorism, epidemics, etc. Section 6.3 refers to Section 8.4, which outlines the claim notification process, so no pre-claim notifications are required.
- **EJCDC C-700 – Section 4.05 and Section 11.09.B:** Section 4.05.C indicates that if the contractor is impacted due to issues that are not the fault of the owner or the engineer, the contractor is entitled to an extension of time as its sole remedy and, per Section 4.05.D.3, such an extension is subject to the provisions of Article 11. Per Section 11.09, the contractor shall submit change proposals to the engineer within 30 days after the start of the event giving rise thereto. The engineer is to render an initial decision on the change proposal within 30 days of proper submission. If contractor does not agree with the engineer's initial decision, it can move to the dispute resolution procedures in Article 12.

## E. Change Order Negotiation Issues

Standard construction contracts typically include procedures to address the scenario where the contractor and the owner cannot agree to the terms of a change order. Here, there is no dispute over whether or not the contractor is entitled to a change order, it is just a dispute over the cost or time impacts of the change. Such provisions typically include an owner directive that mandates that the contractor perform the work, and then notes how the contractor can resolve the cost and time impact issues. If the cost and time negotiations remain unresolved after this process, the contractor can proceed with dispute resolution. The underlined text below represents specific pre-claim requirements.

- **AIA A201 – Section 7.3:** If the contractor and the owner cannot reach an agreement on the change to the contract sum or contract time relating to a change order, the owner can issue a construction change directive that forces the contractor to proceed with the change-related work. Per Section 7.3.6, the contractor is to advise the architect if it disagrees with the method, if any, provided in the construction change directive for determining the proposed adjustment in the contract sum or contract time. The contractor can then move to issue a claim per Article 15.
- **ConsensusDocs 200:** Per Section 8.2, if the contractor and the owner cannot agree to the terms of a change order, the owner can issue an interim directive to the contractor that requires the contractor to perform the subject work. At that point the contractor can move directly to dispute resolution per Article 12. Hence, no pre-claim notifications are required.
- **EJCDC C-700 – Section 11.03:** If the contractor and the owner cannot agree upon the cost and time impacts relating to a change order, the owner can issue a work change directive to the contractor and the contractor shall proceed with the work. Per Section 11.03.B, the contractor shall issue a change proposal to the engineer no later than 30 days after the completion of the work set out in the work change directive. The engineer shall complete its review of the contractor's change proposal within 30 days of receipt. The contractor can then appeal the engineer's decision by filing a claim under Article 12.

## II. Claim Notice Provisions

Once the claimant fulfills the pre-claim requirements, it can initiate the formal claims process by properly noticing the respondent about the claim, if required by contract. Most construction contracts include claim notice procedures and it is important to identify how the notice of a claim can be served, as certain standard contract forms still do not include electronic delivery as an acceptable form of notice.

Notice provisions place the burden on the claimant to give the respondent a heads up on issues that involve impacts that affect the claimant's cost and/or time on the project. Timely notice benefits the respondent because it allows for a timely investigation of the claim issue so sound decisions can be made in a contemporaneous manner. In certain instances, failure to timely notice the respondent may prejudice the respondent.

Some claimants are hesitant to follow contractual claim procedures for fear of the negative effect it might have to the performing party–paying party relationship. Other claimants fear that the issuance of formal claims might cause respondent retaliation. As a result, claimants often bundle claim issues and submit them at the tail end of projects, which can catch respondents by surprise, and this typically leads to unproductive communications and binding dispute resolution. One way to avoid such discomfort is for the parties to review the dispute resolution process up front and periodically during the production phase so both parties are reminded that the defined claim protocol is there to resolve issues early in the dispute resolution process and it serves to prevent claims from snowballing to a point where binding dispute resolution is the only way to achieve resolution. Resolving claims early is typically a benefit to both parties.

Each of the examples below is based on a contractor being the claimant and an owner being the respondent and the two parties have entered a standard contract form and the owner is responsible for the design on the project. The examples cover the notice of claim provision typically required for general claim items, claims regarding a lack of adequate funding, claims related to a failure to make timely payments, and termination by the contractor.

## A.  Notice of Claim Provisions for General Claim Items

Certain standard contract forms require the claimant to issue a formal notice of claim to the respondent prior to issuing complete claim submissions. This makes sense because it typically takes time to calculate cost and time impacts associated with claims. It is standard for notice of claims to be issued by the claimant within 14–21 days of the event giving rise to the claim or first recognition of the event.

### 1.  AIA A201 General Conditions

Per Section 1.6.2 of the AIA A201, the claimant (owner or contractor) shall serve notice of claims in writing and deliver such notices via certified or registered mail, or by courier providing proof of delivery. Per Section 15.1.3 of the A201, for claims where the condition giving rise to the claim is first discovered prior to the expiration of the contractor's one-year warranty period, the claimant (owner or contractor) shall initiate the claims process by providing notice to the initial decision maker, the other party, and the architect. Claims by either party must be initiated within 21 days after the occurrence of the event giving rise to such claim or within 21 days after the claimant first recognizes the condition giving rise to the claim, whichever is later. When the condition giving rise to the claim is first discovered after expiration of the contractor's one-year warranty period, it shall be initiated by notice just to the other party.

---

**Sample Notice Letter that Conforms to the AIA A201**

Via Certified Mail

April 23, 20__

Owner
Address

Re: Project: Project Name

Subject: Section 15.1.3.1 Notice of Claim

---

*(Continued)*

**(Continued)**

Dear Owner/Initial Decision Maker/Architect,

Pursuant to Section 15.1.3.1 of the AIA A201 General Conditions, Contractor hereby submits a formal *Notice of Claim* for the additional scope of work identified on the revised architectural plans that Owner issued to Contractor one week ago, on April 23, 20__. Specifically, Drawing A.2.06 indicates approximately 80 additional feet of wall assembly that was not contemplated in the bid set of plans that the Contract Price is based upon. Thus, this scope is outside of the defined Work. Contractor is calculating cost and time impacts associated with this additional work and will issue a formal Claim for an initial decision per Section 15.2 of the A201 soon. If you have any questions or comments, please contact Contractor at your convenience.

Sincerely,
Contractor

## 2.  ConsensusDocs 200

Per Section 13.5 of the ConsensusDocs 200, the contractor's written notice for claims is effective "upon transmission by any effective means." Thus, notice can be served via electronic mail, US postal, or overnight delivery. Per Section 8.4 of the ConsensusDocs 200, the contractor shall issue a "changes notice" of claim to the owner within 14 days of the occurrence giving rise to the claim or the contractor's first recognition of the condition giving rise to the claim. The contractor shall submit written documentation in support of the claim within 21 days of the notice. The owner shall respond to the claims no later than 14 days after receipt. If the owner does not respond within 14 days or the contractor takes exception to the owner's response, the contractor can trigger dispute resolution per Article 12.

**Sample Notice Letter that Conforms to the ConsensusDocs 200**

Via Certified Mail

April 23, 20__

Owner
Address

Re: Project: Project Name

Subject: Section 8.4 Notice of Claim

Dear Owner,

Pursuant to Section 8.4 of the ConsensusDocs 200 contract, Contractor hereby submits a formal notice of claim for the additional scope of work identified on the revised architectural plans that Owner issued to Contractor one week ago, on April 23, 20__. Specifically, Drawing A.2.06 indicates approximately 80 additional feet of wall assembly that was not contemplated in the bid set of plans that the contract price is based upon. Thus, this scope is outside of the defined Work. Contractor is calculating cost and time impacts associated with this additional work and will issue a formal Claim with supporting documentation within 21 days of this notice of claim. If you have any questions or comments, please contact Contractor at your convenience.

Sincerely,
Contractor

### 3. EJCDC C-700 General Conditions

The EJCDC C-700 general conditions do not have a specific notice requirement for claims. Section 12.01.B notes that contractor shall deliver claims to the owner and the engineer no later than 30 days after the start of the event giving rise thereto. Therefore, unlike the AIA A201 and Consensus-Docs 200, the claims process starts with the submission of the claim itself within 30 days of the event giving rise to the claim, such as a rejected change proposal by the engineer. This forces the contractor to assemble claims in an expedited fashion. Below is a sample cover letter that might accompany a Section 12.01.B claim.

**Sample Claim Cover Letter that Conforms to the EJCDC C-700**

Via Certified Mail

April 23, 20__

Owner
Address

Re: Project: Project Name

*(Continued)*

| |
|---|
| **(Continued)** |
| Subject: Section 12.01.B Claim |
| Dear Owner/Engineer, |
| Pursuant to Section 12.01.B of the EJCDC C-700 General Conditions, Contractor hereby submits a formal claim for the additional scope of work identified on the revised engineering plans that Engineer issued to Contractor one week ago, on April 23, 20__. Specifically, Drawing A.2.06 indicates approximately 80 additional feet of wall assembly that was not contemplated in the bid set of plans that the Contract Price is based upon. Thus, this scope is outside of the defined Work. The following claim includes sections on entitlement, delay, and damages associated with this issue. If you have any questions or comments, please contact Contractor at your convenience. |
| Sincerely,<br>Contractor |

## B. Notice of Claim Provisions for Lack of Evidence of Owner's Financial Arrangements

Most standard contract forms require the owner, upon formal request from the contractor, to request confirmation from the owner of adequate project financing. In the event the owner fails to provide proof of funding, the contractor is not required to start the work and it is entitled to a time extension until such proof is provided.

### 1. AIA A201 General Conditions

Per Section 2.2.1 of the AIA A201, prior to commencement of the work, the contractor may issue a written request to the owner for proof of funding for the project. The contractor has no obligation to commence the work until the owner provides such evidence. If commencement of the work is delayed due to this issue, the contract time shall be extended appropriately.

| |
|---|
| **Sample Request for Proof of Funding Letter that Conforms to the AIA A201**<br>Via Certified Mail<br>April 23, 20__ |

Owner
Address

Re: Project: Project Name

Subject: Section 2.2.1 Proof of Funding Request

Dear Owner,

Pursuant to Section 2.2.1 of the AIA A201 General Conditions, Contractor hereby requests that Owner provide proof of adequate funding for the Project. This information is a condition precedent to Contractor's commencement of the Work. If you have any questions or comments, please contact Contractor at your convenience.

Sincerely,
Contractor

## 2. ConsensusDocs 200

Per Section 4.2 of the ConsensusDocs 200, before commencing the work and thereafter, the contractor may issue a written request to the owner for proof of adequate funding for the project. Evidence of such funding shall be a condition precedent to contractor commencing or continuing the work.

**Sample Request for Proof of Funding Letter that Conforms to the ConsensusDocs 200**
Via Email

April 23, 20__

Owner
Address

Re: Project: Project Name

Subject: Section 4.2 Proof of Funding Request

Dear Owner,

Pursuant to Section 4.2 of the ConsensusDocs 200, Contractor hereby requests that Owner provide proof of adequate funding for the Project.

*(Continued)*

---

**(Continued)**

This information is a condition precedent to Contractor's commencement of the Work. If you have any questions or comments, please contact Contractor at your convenience.

Sincerely,
Contractor

---

### 3. EJCDC C-700 General Conditions

Per Section 9.11 of the EJCDC C-700, upon request of the contractor, the owner shall furnish the contractor with evidence of funding for the project. Unlike the A201 and the ConsensusDocs 200, the C-700 does not mention that the contractor should hold up the commencement of work until the owner has complied with this request.

---

**Sample Request for Proof of Funding Letter that Conforms to the EJCDC C-700**

Via Hand Delivery

April 23, 20__

Owner
Address

Re: Project: Project Name

Subject: Section 9.11 Proof of Funding Request

Dear Owner,

Pursuant to Section 9.11 of the EJCDC C-700 General Conditions, Contractor hereby requests that Owner provide proof of adequate funding for the Project. If you have any questions or comments, please contact Contractor at your convenience.

Sincerely,
Contractor

---

## C. Notice of Claim Provisions for the Owner's Failure to Make Timely Payment to the Contractor

If the owner fails to make timely payment to the contractor, most standard contract forms allow the contractor to stop work after proper notice to owner until full payment is received, plus interest.

### 1. AIA A201 General Conditions

Per Section 9.7 of the AIA A201, if the architect fails to certify the contractor's payment application within seven days of receipt or if the owner fails to fund the contractor's certified payment application within seven days from when required by contract, then the contractor may, upon seven days' notice to the owner and the architect, stop the work until the issue is cured. The contract time shall be extended appropriately, and the contract sum shall be increased to account for the contractor's reasonable costs of shutdown, delay, and start-up. In addition, the contractor is entitled to interest as provided for in the contract documents.

---

**Seven-Day Stop Work Notice per the AIA A201**

Via Certified Mail

April 23, 20_

Owner and Architect
Address

Re: Project: Project Name

Subject: Section 9.7 – Late Payment Application Certification Notice/ Stop Work Notice

Dear Owner and Architect,

Pursuant to Section 9.7 of the AIA A201 General Conditions, Contractor hereby notices Owner and Architect of Architect's failure to certify Contractor's Payment Application No. 6 within seven days as required by Section 9.4.1 and Architect has further failed to certified this application within an additional seven days thereafter. Thus, Contractor

---

*(Continued)*

**(Continued)**

intends to stop work on the Project if certification is not made within seven days of that date of this notice. If you have any questions or comments, please contact Contractor at your convenience.

Sincerely,
Contractor

## 2. ConsensusDocs 200

Per Section 9.5 of ConsensusDocs 200, if the contractor does not receive payment from the owner, through no fault of the contractor, within seven days after payment is due, then the contractor can provide the owner a seven-day notice and may stop work until the owner pays the contractor the full amount due, plus interest.

**Seven-Day Stop Work Notice per the ConsensusDocs 200**

Via Email

April 23, 20__

Owner
Address

Re: Project: Project Name

Subject: Section 9.5 – 7-Day Stop Work Notice for Lack of Payment

Dear Owner,

Pursuant to Section 9.5 of the ConsensusDocs 200, Contractor hereby notices Owner of its failure to fund Contractor's approved Payment Application No. 12 within 30 days per the Agreement plus an additional seven days. Hence, Contractor hereby provides notice to the Owner that Contractor will stop work on the Project if Owner fails to fund this payment application within seven days of the date of this letter. If you have any questions or comments, please contact Contractor at your convenience.

Sincerely,
Contractor

### 3. EJCDC C-700 General Conditions

Per Section 16.04.B of the EJCDC C-700, if the engineer fails to act on the contractor's payment application within 30 days after submission, or the owner fails to pay the contractor within 30 days after it becomes due, the contractor may, after seven days written notice to the owner and the engineer, stop the work until full payment is made, plus interest.

---

**Seven-Day Stop Work Notice per the EJCDC C-700**

Via Hand Delivery

April 23, 20__

Owner and Engineer
Address

Re: Project: Project Name

Subject: Section 16.04.B – 7-Day Stop Work Notice for Lack of Certification

Dear Owner and Engineer,

Pursuant to Section 16.04.B of the EJCDC C-700, Contractor hereby notices Owner and Engineer of Engineer's failure to take action on Contractor's Payment Application No. 7 within 30 days per the Agreement. Hence, Contractor hereby provides notice to Owner and Engineer that Contractor will stop work on the Project if Engineer fails to act on this payment application within seven days of the date of this letter. If you have any questions or comments, please contact Contractor at your convenience.

Sincerely,
Contractor

---

## D. Notice of Termination Provisions by the Contractor to the Owner

While most contract forms include a provision whereby the contractor can terminate the owner, this remedy is rarely exercised. Regardless, it is important for contractors to understand the notice requirements and scenarios where it can terminate the owner.

### 1. AIA A201 General Conditions

Per Section 14.1 of the AIA A201, the contractor can terminate the owner, upon seven days' written notice to the owner and the architect, if any of the following scenarios exist, through no fault of the contractor:

- If the work is stopped for a period of 30 consecutive days because:
  - an order of a court or other public authority that requires all work to be stopped;
  - an act of government requires all work to be stopped;
  - the architect has not certified the contractor's payment application or the owner has not made payment to the contractor within the time stated in the contract documents; or
  - the owner has failed to furnish the contractor proof of funding per Section 2.2.
- if the owner repeatedly suspends, delays, or interrupts the entire work and it causes a delay, in aggregate, of more than 100 percent of the total number of days scheduled for completion, or 120 days in any 365-day period, whichever is less.
- if work is stopped for 60 consecutive days because the owner has repeatedly failed to fulfill the owner's obligations under the contract documents with respect to matters important to the progress of work.

Upon proper termination by the contractor, the contractor is entitled to recover payment for all work executed, as well as reasonable overhead and profit on work not executed, and costs incurred as a result of such termination.

---

**Notice of Contractor Termination of Owner per the AIA A201**

Via Certified Mail and Hand Delivery

April 23, 20__

Owner and Architect
Address

Re: Project: Project Name

Subject: Section 14.1 – 7-Day Notice of Termination

Dear Owner and Architect,

Pursuant to Section 14.1 of the AIA A201 General Conditions, Contractor hereby notices Owner and Architect of its intention to terminate the Contractor's agreement with the Owner within seven (7) days of the date of this letter. The reason for this notice of termination is a result of the court order that Contractor received on February 1, 20__, which is well over 30 consecutive days ago, that prevents Contractor from continuing the Work on the Project for reasons that have nothing to do with the Contractor. The court order, which is attached herein, relates to the Owner's dispute with a neighboring property owner. If you have any questions or comments, please contact Contractor at your convenience.

Sincerely,
Contractor

## 2. ConsensusDocs 200

Per Section 11.5 of the ConsensusDocs 200, the contractor can terminate the agreement, if the contractor issues the owner a seven-day written notice of termination through no fault of contractor, and:

- the work is stopped for a 30-day period due to one of the following scenarios:
  - court order or order of other governmental authority;
  - governmental act during which materials are not available; or
  - suspension by owner for convenience per Section 11.1.
- the owner fails to cure one of the following issues within three days upon receipt of the contractor's notice of termination:
  - fails to provide the contractor proof of funding per Section 4.2;
  - assigns the agreement upon the contractor's reasonable rejection;
  - fails to pay the contractor per the terms of the agreement and the contractor has stopped work per Section 9.5; or
  - otherwise materially breaches the agreement.

Upon proper termination by the contractor, the contractor is entitled to recover payment for all work executed and for any proven loss, cost, or expense in connection with the work, including all demobilization costs plus reasonable overhead and profit on work not performed.

**Notice of Contractor Termination of Owner per the ConsensusDocs 200**
Via Certified Mail and Hand Delivery
April 23, 20__

*(Continued)*

**(Continued)**

Owner
Address

Re: Project: Project Name

Subject: Section 11.5 – 7-Day Notice of Termination

Dear Owner,

Pursuant to Section 11.5 of the ConsensusDocs 200, Contractor hereby notices Owner of its intention to terminate the Contractor's agreement with the Owner within seven (7) days of the date of this letter. The reason for this notice of termination is a result of the court order that Contractor received on February 1, 20__, which is well over 30 consecutive days ago, that prevents Contractor from continuing the Work on the Project for reasons that have nothing to do with the Contractor. The court order, which is attached herein, relates to the Owner's dispute with a neighboring property owner. If you have any questions or comments, please contact Contractor at your convenience.

Sincerely,
Contractor

### 3.   EJCDC C-700 General Conditions

Per Section 16.04.A of the EJCDC C-700, the contractor can terminate the agreement, upon issuance of a seven-day notice to the owner and the engineer and provided the owner or the engineer do not remedy the issue within such time, if one of the following scenarios exists through no fault of the contractor:

- if work is suspended for more than 90 consecutive days by the owner, court order, or order from other public authority;
- if the engineer fails to act on a contractor payment application within 30 days from submission; or
- if the owner fails to fund the contractor's payment application within 30 days from when it is due.

In the event of a proper termination by the contractor, the contractor is entitled to recover from the owner costs for completed and uncompleted work plus fair and reasonable sums for overhead and profit, as well as expenses directly attributable to termination.

---

**Notice of Contractor Termination of Owner per the EJCDC C-700**

Via Certified Mail and Hand Delivery

April 23, 20__

Owner and Engineer
Address

Re: Project: Project Name

Subject: Section 16.04.A – 7-Day Notice of Termination

Dear Owner and Architect,

Pursuant to Section 16.04.A of the EJCDC C-700 General Conditions, Contractor hereby notices Owner and Engineer of its intention to terminate the Contractor's agreement with the Owner within seven (7) days of the date of this letter. The reason for this notice of termination is a result of the court order that Contractor received on November 15, 20__, which is well over 90 consecutive days ago, that prevents Contractor from continuing the Work on the Project for reasons that have nothing to do with the Contractor. The court order, which is attached herein, relates to the Owner's dispute with a neighboring property owner. If you have any questions or comments, please contact Contractor at your convenience.

Sincerely,
Contractor

---

# 5

## Step 4: Establish Entitlement

## I. Introduction

Proper construction claims have three major parts: (1) entitlement; (2) delay, if applicable; and (3) damages. Claimants often take a cursory approach to proving entitlement and focus primarily on the calculation of delay and damages related to claims. However, if the claimant fails to establish entitlement for a claim, it generally cannot recover time or damages. Thus, claimants should not gloss over entitlement. The entitlement section of a claim tells the story of the dispute and links the cause to the effect of the issue. Once the entitlement threshold is properly cleared, the claimant can move on to the delay analysis, if applicable, and then to damages. Creating an outline and identifying the information that is needed to prove entitlement keeps claimants organized and it makes the claim process more efficient.

For entitlement outlines, I favor the strategy taught in many law schools called "CRAC," which stands for Conclusion, Rule, Analysis, and Conclusion. CRAC is often used to organize legal arguments in essays, memos, or briefs in court. Far too often, construction claims are presented like mystery novels, where the reader doesn't know what the claimant is asking for until the end of the last page of the claim narrative. While this strategy might prove interesting, it is very ineffective because the respondent wants to know what the claimant is asking up front, so it can better understand the rules laid out by the contract documents and how the facts compare to the rules, so the reader can readily determine the merits of the claim and the potential for recovery. Waiting until the end of an entitlement narrative to draw conclusions often forces the reader to reread the narrative—an activity most respondents detest. Restating the conclusion at the end of the entitlement narrative serves to reinforce the claimant's position. Note that the ultimate headers of the entitlement narrative do not need to read "Conclusion," "Rule," "Analysis," and "Conclusion," but the flow of information should match this organizational structure.

Entitlement discussions often involve a review of complicated contract provisions and technical subjects related to the contract documents and unanticipated conditions. The drafter of an entitlement narrative must do his or her best to convey relevant concepts as if they were being explained at a layperson level, which is easier said than done. If the entitlement narrative is overly technical, there is a good chance it will not be understood by the respondent, or the respondent will not spend the time to understand the subject matter. Moreover, if the claim is later decided in binding dispute resolution where the finder of fact has little or no experience with the subject matter, which is often the case in jury trials, the subject will need to be distilled into simple terms at that time. so it is better to communicate in a clear and concise manner up front. College architecture, engineering, and construction programs spend little time on technical writing, and many architecture, engineering, and construction (AEC) professionals struggle with this subject, and that is why many firms hand over claim writing to attorneys and consultants, which can prolong the claims administration process. Technical writing is a learned skill and the more a person writes, the better he or she will be at entitlement narratives. My advice for AEC firms is to encourage staff members to take college-level classes in technical writing, particularly because online learning is readily accessible and course offerings are plentiful.

The following sections identify common types of claims for various claimants and provide entitlement outlines for such claims in a CRAC format.

## II. Typical Contractor Claims Against Owners

The most common types of claims that contractors assert against owners relate to design issues, administrative issues, owner performance issues, force majeure issues, and change order negotiation issues. The following discussions assume that the claimant is the contractor and the owner-contractor agreement is an AIA document that incorporates the A201 general conditions.

## A.  Owner Design Issues

Entitlement for design-related claims generally involves the identification of a changed condition through a comparison of the work outlined in the contract documents and a changed condition, which affects the contractor's contract sum and/or contract time. The change might be an unanticipated subsurface condition or changed work on an updated set of plans or request for information (RFI) response. The agreement between the parties should detail the contract documents that the contractor is bound to, so this exercise is often straightforward.

### 1.  Differing Site Condition Claims

Differing site condition claims often relate to construction contracts that involve earthwork, demolition, or improvements to existing structures. Owners typically take soil borings, material samples, or perform localized destructive testing to reasonably anticipate what underlying conditions consist of so contractors have a basis to bid related divisions of work. When it turns out that the owner's information does not properly convey what is ultimately in place, the contractor is generally entitled to recover additional cost and time impacts resultant from this type of unforeseen condition.

Differing site condition claims are broken into two categories: Type 1 and Type 2. Type 1 differing site condition claims are more common than Type 2 claims, and they involve a subsurface or concealed physical condition that differs materially from those indicated in the contract documents. For instance, a Type 1 differing site condition may exist if the contractor encounters rock during excavation when the contract documents explicitly indicate sand. A Type 2 changed condition is an unexpected physical condition that is of an unusual nature and differs materially from the types of conditions ordinarily encountered and generally recognized in the physical location of the work. Type 2 conditions are typically asserted when the contract is silent regarding subsurface conditions. Type 2 differing site conditions are less common and harder to prove than Type 1 claims. An example of a Type 2 differing site condition is when a contractor encounters a below-grade structure in an area that appeared only to contain soil.

**<u>Hypothetical Type 1 Differing Site Condition Claim</u>:** The owner retained the contractor to construct a commercial building under an AIA A101 agreement (design-bid-build with a stipulated sum) that incorporated the AIA A201 general conditions. The owner previously retained the geotechnical engineer to obtain borings on the site in order to identify the subsurface conditions and to recommend an appropriate type of foundation system. The contractor based its earthwork and underground utilities estimate upon this geotechnical report, which is a named contract document. The geotechnical report notes that subsurface conditions are silty sand material to a depth of approximately 25 feet over the entire project site. The contractor reasonably relied on this information.

Shortly after the contractor started its 15-foot excavation in order to install the foundation system and basement slab on ground work, it encountered large rock outcroppings at a depth of 8 feet that cover approximately 20% of the project site. As a result, the rock outcroppings must be removed by blasting, which the contractor did not anticipate or account for in its cost estimate or project schedule.

**Sample Entitlement Outline:**

A) **Conclusion**

The contractor hereby requests an initial decision from the initial decision maker (IDM) regarding the contractor's Type 1 differing site condition on the project. Specifically, the contractor seeks an extension to the contract time of XX calendar days and an increase in the contract sum of $XX.

B) **Rule**

Article 1 of the A101 defines the contract documents for the project; the A201 and the owner's geotechnical report are listed as contract documents. Section 3.7.4 of the A201 requires the contractor to notice the owner and architect within 14 days of identifying a differing site condition. The architect must then promptly investigate this issue and either deny the claim or recommend an equitable adjustment to the contract sum and/or time.

In the event the architect rejects the differing site condition claim, Section 15.1.3.1 requires the contractor to notice the owner, the architect, and the IDM within 21 days of the occurrence of the event giving rise to the claim or recognition of the event giving rise to the claim.

Per Section 15.2.1, the contractor shall then assemble and then refer the claim to the IDM for an initial decision. If the IDM rejects the claim, the contractor may trigger mediation and binding dispute resolution according to Article 15 requirements.

C) **Analysis**

The owner's geotechnical report, which is a contract document under Article 1 of the A101, includes 25 subsurface borings throughout the site. The borings all identify the subsurface material to be silty-sand material to a depth of approximately 25 feet. A copy of this geotechnical report is attached herein. The contractor reasonably relied on this design information to estimate the excavation of the building foundation system.

Shortly after the contractor commenced its foundation excavation work, it encountered a large rock outcropping at a depth of approximately 8 feet. Based on exploratory work performed by the contractor, it appears that the outcropping covers approximately 20% of the foundation area. Accordingly, the contractor will need to blast this rock in order to excavate to the bottom of footing depth. Because the contract documents clearly identify silty sand as the subsurface condition to a depth of 25 feet, and because the contractor could not have reasonably discovered this differing site condition through a pre-bid site investigation, and the contractor encountered a significant rock outcropping at 8 feet, this claim is a Type 1 differing site condition. Photographs of the exposed rock outcroppings are included herein, along with a legend that identifies the location of the rock.

Per Section 3.7.4 of the A201, the contractor notified the owner and architect of this differing site condition well within 14 days upon observance. The architect subsequently reviewed this issue and formally denied the claim. As a result, the contractor later noticed the owner, the architect, and the IDM of this claim within 21 days of the architect's rejection of the contractor's differing site condition request. Copies of the architect's rejection and the contractor's notice of claim are attached herein. In addition, the delay section of this claim outlines the time impacts associated with the differing site condition. The damages section of this claim identifies the costs impacts of this differing site condition.

Per Section 15.2.1, the contractor hereby files this claim to the IDM for an initial decision. In the event that the IDM rejects this claim, the contractor will move forward with mediation and binding dispute resolution per Article 15 of the A201.

D) **Conclusion**

The contractor seeks an initial decision from the IDM for the contractor's Type 1 differing site condition claim per the terms of the A201. Specifically, the contractor seeks an adjustment in the contract time of XX calendar days and an increase in the contract sum of $XX.

**Hypothetical Type 2 Differing Site Condition Claim:** The owner retained the contractor to construct an office building in Miami, FL, under an AIA A102 agreement (design-bid-build with a guaranteed maximum price (GMP)) that incorporated the AIA A201 general conditions. Work on the project includes running underground utility lines to the building. The contract documents are silent on the geotechnical conditions where the utility lines run. The contractor conducted a visual investigation of the site prior to bidding the work but did not perform any borings, as that was not a requirement per the request for proposal (RFP). Based on the contractor's experience in working throughout the greater Miami area for the past 40 years, it assumed the subsurface conditions where the utilities were located were sand.

Shortly after the contractor started its excavation for the underground utility work, it encountered large boulders that required the use of much larger equipment than the contractor had anticipated in its bid, which impacted the contractor's cost and schedule of this work.

**Sample Entitlement Outline:**

A) **Conclusion**

The contractor hereby requests an initial decision from the initial decision maker (IDM) on the contractor's Type 2 differing site condition claim relating to unusual subsurface conditions where the underground utilities are located at the project. Specifically, the contractor seeks an extension to the contract time of XX calendar days and an increase in the contract sum of $XX.

B) **Rule**

Article 1 of the A102 lists the contract documents that the contractor is bound to—the A201 is listed as a contract document. Section 3.7.4 of the A201 requires the contractor to notice the owner and architect within 14 days of identifying a differing site condition. The architect must then promptly investigate this issue and either deny the claim or recommend an equitable adjustment to the contract sum and/or time.

In the event the architect rejects the differing site condition claim, Section 15.1.3.1 requires the contractor to notice the owner, the architect, and the initial decision maker within 21 days of the occurrence of the event giving rise to the claim or recognition of the event giving rise to the claim.

Per Section 15.2.1, the contractor shall then assemble and then refer the claim to the IDM for an initial decision. If the IDM rejects the claim, the contractor may trigger mediation and binding dispute resolution according to Article 15 requirements.

C) **Analysis**

The contract documents, as detailed in Article 1 of the A102 agreement, do not include a geotechnical report that identifies the subsurface conditions where the underground utility lines are located. The contractor reasonably based its estimate of the excavation work associated with the underground utility work upon typical geotechnical conditions that the contractor has encountered in the Greater Miami area over the past 40 years. The contractor further supports this assumption with three geotechnical reports on projects that the contractor has worked on within close proximity to the subject project, each of which denote sand as the typical subsurface material. Copies of these reports are included herein.

Shortly after the contractor commenced its utility excavation work, it encountered large boulders that could not be removed with the equipment being used by the contractor. Photographs of these boulders are included herein. Per Section 3.7.4 of the A201, the contractor notified the owner and the architect of this Type 2 differing site condition within 14 days upon observance. The architect subsequently reviewed this issue and formally denied the claim. As a result, the contractor later noticed the owner, the architect, and the IDM of this claim within 21 days of the architect's rejection of the contractor's differing site condition request. Copies of the architect's rejection and the contractor's notice of claim are attached herein. Based on the architect's rejection, the contractor has mobilized larger equipment to remove these boulders in order to mitigate the impacts of this claim.

Per Section 15.2.1 of the A201, the contractor hereby files this claim with the IDM, with copies to the owner and the architect, for an initial decision. The time and cost impacts associated with this claim are detailed in the delay and damages section of this report, respectively.

D) **Conclusion**

The contractor hereby requests an initial decision from the IDM for the contractor's Type 2 differing site condition claim. Specifically, the contractor seeks an extension to the contract time of XX calendar days and an increase in the contract sum of $XX.

## 2. Design Additions/Design Changes

It is common for designers to update contract documents after execution of the owner-contractor agreement, particularly if the delivery method has an overlap between the design and contractor phases of the project. When this occurs and the updated plans include added or changed work items that were not contemplated in the contract documents that the contractor used to bid for the project, contractors are entitled to recover the cost and time impacts the additional work has on the project.

**Hypothetical Added Scope Claim Scenario:** The owner retained the contractor to construct a seven-story multi-family building with underground parking. The owner-contractor agreement is an AIA A133 agreement (Construction Manager at risk) that incorporates the AIA A201 general conditions. After the owner and the contractor executed the guaranteed maximum price amendment, the designer materially revised the architectural plans in the basement area of the project. Specifically, the new plans note additional wall assemblies not found on the set of plans that the GMP is based upon.

**Sample Entitlement Outline:**

A) **Conclusion**

The contractor hereby requests an initial decision from the IDM for its added scope of work claim relating to the additional wall assemblies in the basement area of the project that the designer added in its most recent plan revisions. This represents additional work that will increase the contract sum by $XX.

B) **Rule**

Per Section 3.2.3.1 of the A133, the GMP amendment should list the drawings and specifications that the GMP is based upon. Section 3.2.2 of the A133 notes that unanticipated change to the contract documents after the GMP amendment is executed is not contemplated within the contractor's GMP and impacts to such changes should be incorporated into the contractor's scope of work via a change order.

If the owner and the architect neglect to issue a change order to the contractor for unanticipated design changes after execution of the GMP amendment, the contractor has the ability to notice the owner, the architect, and the IDM, per Section 15.1.3.1 of the A201 of such a claim within 21 days of the occurrence of the event giving rise to the claim or recognition of the event giving rise to the claim.

Per Section 15.2.1, the contractor shall then assemble and refer the claim to the IDM for an initial decision. If the IDM rejects the claim, the contractor may trigger mediation and binding dispute resolution according to Article 15 requirements.

C) **Analysis**

The GMP amendment identifies the contract drawings that are dated January 15th. Per Section 3.2.3.1 of the A133, this is the set of plans that the contractor is bound to. On August 20th, approximately seven months after the architect issued its January plan set, the architect reissued several plans relating to the basement area of the project.

Specifically, the GMP drawings include 250 linear feet of interior wall assembly in the basement whereas the updated architectural plans include 550 linear feet of wall assembly in the basement; thus, the architect added an additional 300 linear feet of wall assembly that fall outside of the contractor's scope of work. The additional wall assembly work increases the cost to the following divisions of work: framing, rough MEP, drywall, wall finishes, and MEP trim.

Per Section 3.2.2 of the A133, any unanticipated changes to the GMP amendment documents should be incorporated into the contractor's scope of work via change order. Within 21 days of receipt of the August plan set, the contractor noticed the owner, the architect, and the IDM of a claim relating to this additional work, as required under Section 15.1.3.1 of the A201. Weeks have gone by without the owner and the architect issuing a change order to the contractor for this additional work, so the contractor hereby requests an initial decision from the IDM per Section 15.2.1 of the A201.

If the IDM rejects the claim, the contractor may trigger mediation and binding dispute resolution according to Article 15 requirements.

D) **Conclusion**

In sum, the contractor is entitled to impacts to the contractor's contract sum of $XX caused by the additional wall assembly work that the designer added to the basement area of the project in its recent August set of plans. Accordingly, the contractor seeks an initial decision from the IDM.

## B. Administration Issues

Establishing entitlement for maladministration impacts includes a comparison of the owner's timelines requirements noted in the owner-contractor agreement or industry standard guidelines against the owner's actual performance. Maladministration issues often involve payment, funding, requests for information (RFIs), submittals, change orders, and inspections. It is unusual for one delinquent administrative issue to cause a significant cost and/or time impact on the project. However, it is common for a consistent pattern of delinquent behavior from the owner or its representatives to give rise to a maladministration impact.

**Hypothetical Owner Maladministration Claim: Late Payments:**

The owner retained the contractor to construct an industrial warehouse project. The owner-contractor agreement is an AIA A133 agreement (construction manager (CM) at risk) that incorporates the AIA A201 general conditions. During the construction phase of the project, the owner continually makes late payments to the contractor, which violates the terms of the contract.

**Sample Entitlement Outline:**

A) **Conclusion**

The contractor seeks an initial decision from the IDM for impacts related to the owner's continual breach of the payment provisions of the owner-contractor agreement over the past six months. As a result of the owner's actions, the contractor seeks approval of this claim and a change order to extend the contract time by XX days and an increase in the contract sum by $XX.

B) **Rule**

Section 7.1.3 of the AIA A133 agreement notes that the owner is to pay the contractor within 30 days of the contractor's submission of its application for progress payment to the architect, so long as the architect certifies the application within the noted time period.

Section 9.6.1 of the A201 notes that after the architect's certification of the contractor's payment application, the owner shall make payment within the time provided in the agreement. Per Section 8.3.1 of the A201, if the contractor is delayed by an act or neglect of the owner or the architect, the architect shall determine what time, if any, is owed to the contractor. If the contractor disagrees with architect's position, it may make a claim per Article 15.

Section 15.1.3.1 of the A201 requires the contractor to notice the owner, the IDM, and the architect within 21 days of the occurrence of the event giving rise to the claim or recognition of the event giving rise to the claim. Post 15.1.3.1 notice, the contractor shall refer the claim to the IDM for initial decision per Section 15.2.1.

C) **Analysis**

Section 7.1.3 of the AIA A133 agreement notes that the owner is to pay the contractor within 30 days of the contractor's submission of its application for progress payment to the architect, so long as the architect certifies the application within the noted time period.

Section 9.6.1 of the A201 notes that after the architect's certification of the contractor's payment application, the owner shall make payment within the time provided in the agreement. Per Section 8.3.1 of the A201, if the contractor is delayed by an act or neglect of the owner or the architect, the architect shall determine what time, if any, is owed to the contractor. If the contractor disagrees with architect's position, it may make a claim per Article 15.

To date, the contractor has issued six payment applications. The contractor issued each application to the architect by the first of each month. The architect has reviewed and certified each of the six applications by the 15th of each month. Thus, per Section 7.1.3 of the A133, the owner was required to fund each of these applications within 30 days of submission, or by the end of each month. This payment timing requirement is further noted in Section 9.6.1 of the A201. However, the owner has paid the contractor, on average, 60 days after submission, or twice as long as required by contract. Attached to this report is a table that illustrates the owner's delinquent payment turn for payment applications 1–6. The contractor has continually noticed the owner, the architect, and the IDM of these administrative breaches per Section 15.1.3.1 of the A201.

As a result of this slow payment administration, many of the contractor's subcontractors have refused to add the necessary workers on the project and several have threated to suspend the work, which is a remedy the subcontractors possess under the A401 subcontract agreements. As defined in the delay section of this claim, the owner's actions have caused a XX calendar day critical path impact to the project schedule. As detailed in the damages section of this report, the contractor seeks delay damages to the amount of $XX that corresponds with this delay amount.

Per Section 15.2.1 of the A201, the contractor seeks an initial decision from the IDM regarding this impact. If the IDM rejects this claim, the contractor will trigger mediation and, if mediation does not result in settlement, arbitration per the terms of Article 15 of the A201.

D) **Conclusion**

The contractor seeks an initial decision from the IDM on the contractor's claim for impacts relating to the owner's maladministration on the project, as detailed here. The contractor seeks an extension in the contract time by XX calendar days and an increase in the contract sum by $XX.

**Hypothetical Owner Maladministration Claim: Late Submittal Reviews:** The owner retained the contractor to construct a high-rise condominium project. The owner-contractor agreement is an AIA A102 agreement (design-bid-build with a GMP) that incorporates the AIA A201 general conditions. During the construction phase of the project, the owner, through its architect, is continually late in responding to submittals, which has caused trade stacking and delays to the project.

**Sample Entitlement Outline:**

A) **Conclusion**

The contractor seeks an initial decision from the IDM for impacts that the contractor has sustained related to the owner's continual breach, through its architect, of the submittal provisions of the owner-contractor agreement. This continued maladministration has caused a critical path delay to the project. As a result, the contractor seeks an initial decision from the IDM regarding this claim, which includes an equitable adjustment in the contract time of XX days and an increase in the contract sum of $XX for delay damages.

B) **Rule**

Section 3.10.2 of the A201 notes that the contractor shall submit a submittal schedule for the architect's approval. If the contractor fails to submit a submittal schedule, or fails to provide submittals in accordance with the approved submittal schedule, the contractor shall not be entitled to any increase in contract sum or contract time based on the time the architect spends in reviewing submittals. Section 4.2.7 of the A201 indicates that the owner's architect shall return the contractor's submittals per the contractor's approved submittal schedule or in the absence of an approved submittal schedule, with reasonable promptness.

Per Section 8.3.1 of the A201, if the contractor is delayed by an act or neglect of the owner or the architect, the architect shall decide what time, if any, is owed to the contractor. If the contractor disagrees with the architect's position, it may make a claim per Article 15.

Section 15.1.3.1 of the A201 requires the contractor to notice the owner, the initial decision maker, and the architect within 21 days of the occurrence of the event giving rise to the claim or recognition of the event giving rise to the claim. Section 15.2.1 states that the contractor shall refer claims to the initial decision maker for a decision, with a copy to the owner and the architect.

C) **Analysis**

Shortly after the execution of the A102 agreement, the contractor issued its proposed submittal schedule for approval. After incorporating the architect's feedback from this submission, the architect approved the updated schedule. Thus, per Section 3.10.2 of the A201, the contractor can make claims for impacts caused by the architect's failure to adhere to the approved submittal schedule. Section 4.2.7 of the A201 confirms the architect's duty to respond to submittals according to the approved submittal schedule. Note that the contractor cannot install unapproved materials into the work, so timely submittal turns are important to the flow of the overall project.

To date, the contractor has met all submission dates of the approved submittal schedule (85 submittals thus far); however, the architect has not met its required response dates on 72 of the submittals. The architect's untimely review has caused a critical path delay on the project, as detailed in the delay section of this report. The architect is aware of these delays but per Section 8.3.1, it has yet to award the contractor additional time. Submittals are the topic of discussion in every weekly OAC meeting, so the architect and the owner have been aware of this issue for some time.

Moreover, per Section 15.1.3.1 of the A201, the contractor has noticed the owner, the initial decision maker, and the architect of this systemic issue. Per Section 15.2.1, the contractor requests an initial decision from the IDM on this claim.

D) **Conclusion**

The owner, through the architect, has maladministered the submittal process for the project and this maladministration has caused a critical path delay on the project. The delay section of this claim includes a forensic schedule analysis and the damages section of this claim provides support for the delay damages related to this delay. The contractor seeks an initial decision from the IDM regarding the contractor's time extension request of XX calendar days and an increase to the contractor sum of $XX for delay damages. If the IDM rejects this claim, the contractor will trigger mediation and, if mediation does not result in settlement, arbitration per the terms of Article 15 of the A201.

**Hypothetical Owner Maladministration Claim: Late RFI Responses:** The owner retained the contractor to construct a mid-rise office building project. The owner-contractor agreement is an AIA A133 agreement (CM at risk with a GMP) that incorporates the AIA A201 general conditions. During the construction phase of the project, the owner, through its architect, persistently failed to address the contractor's requests for information (RFI) in a timely manner, which caused downtime for certain crews and delays to the critical path of the project. As a result, the contractor seeks an increase to its contract time and contract sum.

**Sample Entitlement Outline:**
### A) Conclusion
The contractor seeks an initial decision from the IDM for impacts that the contractor has sustained related to the owner's continual breach, through its architect, of the RFI submittal provisions of the owner-contractor agreement. This continued maladministration has caused a delay to the critical path of the project. As a result, the contractor seeks an initial decision from the IDM regarding this claim, which includes an equitable adjustment in the contract time of XX days and an increase in the contract sum of $XX for delay damages.
### B) Rule
Per Article 2.1 of the A133, the A201 is a contract document and per Article 2.3.2, it is applicable during the entire construction phase of the project. Section 3.6.4.4 of the A201 notes that the architect will review and respond to the contractor's requests for information about the contract documents within any time limits agreed upon or otherwise with reasonable promptness and, if appropriate, the architect shall prepare and issue supplemental drawings and specifications to the requests for information.

While "reasonable promptness" is an undefined term in the contract documents, several studies have been conducted by construction professionals to define what reasonable promptness means. According to page eight of Navigant Consulting's *Impact & Control of RFIs on Construction Projects* White Paper, the average reply time for RFIs is 10 days based on a review of approximately 1,400 projects.

Section 8.3.1 notes that if the contractor is delayed by an act or neglect of the owner or the architect, the architect shall decide what time, if any, is owed to the contractor. If the contractor disagrees with the architect's position, it may make a claim per Article 15.

Section 15.1.3.1 of the A201 requires the contractor to notice the owner, the architect, and the IDM within 21 days of the occurrence of the event giving rise to the claim or recognition of the event giving rise to the claim. Section 15.2.1 mandates that the contractor shall refer claims to the initial decision-maker for a decision.

C) **Analysis**

To date, the contractor has issued 52 RFIs and the architect's average reply time is 28 calendar days. Below is a table that sets forth the details of each RFI. While the contract is silent on the number of days that the owner must reply to RFIs, Section 3.6.4.4 of the A201 requires the architect to review RFIs with reasonable promptness. Per an industry study from Navigant that profiled over 1,400 construction projects, the average response time for an RFI is 10 calendar days. Thus, 10 calendar days sets the industry standard and the architect has therefore been delinquent in its reply by an average of 18 days (28 calendar days minus 10 calendar days).

The contractor has reminded the architect and the owner of the need for reasonable promptness in RFI turns during the past five OAC meetings. Copies of the meeting minutes are included herein, which highlight these RFI discussions. Although the architect is aware of this issue, it has yet to grant the contractor an adjustment in the contract time as required by Section 8.3.1 of the A201. Consequently, the contractor previously noticed the owner, the architect, and the IDM of a claim for these delays pursuant to Section 15.1.3.1 of the A201.

The delay section of this claim details the impact that this RFI process has had on the project schedule. Specifically, the forensic schedule analysis confirms that the architect's RFI maladministration caused a XX calendar day impact to the substantial completion date of the project schedule. Hence, the contractor hereby requests an adjustment to the contract time of this amount, as well as related extended general conditions in the amount of $XX that is detailed in the damages section of this claim.

D) **Conclusion**

Per Section 15.2.1 of the A201, the contractor hereby requests an initial decision from the IDM on this claim related to the architect's maladministration of the RFI process. To date, this maladministration has caused a XX calendar day delay to the critical path of the project. The contractor requests a modification of the contract time of this number of days and an increase in the contract sum of $XX for related delay damages. If the IDM rejects this claim, the contractor will trigger mediation and, if mediation does not result in settlement, arbitration per the terms of Article 15 of the A201.

## C. Owner Performance Issues

It is not uncommon for owners to undertake certain procurement and performance duties on construction projects. Typical examples include furniture, fixtures, and equipment (FF&E) and certain finish work. When an owner's performance issues interfere with the contractor's work, contractor claims often manifest. Proving owner performance issues is generally done through identifying when the owner's or owner's separate contractor's performance is required against when it actually took place. Key documents often include the baseline schedule, schedule updates the as-built schedule, daily reports, correspondence, and photographs.

**<u>Hypothetical Owner Performance Claim Issue: Delays by Separate Contractor</u>:** The owner retained the contractor to construct a hotel project. The owner-contractor agreement is an A101 agreement (design-bid-build with a stipulated sum) that incorporates the A201 general conditions. The owner also retains a separate contractor to complete demolition of an existing structure before the contractor can commence the foundation work. Shortly after the owner issued the contractor a notice to proceed, the contractor mobilized to the project but was prevented from starting its excavation work because the owner's separate contractor failed to complete its work on time, which caused a delay to the start of contractor's excavation work.

**<u>Sample Entitlement Outline:</u>**
  A) **Conclusion**

  The contractor seeks an initial decision from the IDM for impacts that the contractor has sustained related to the delay caused by the owner's separate demolition contractor, which failed to complete the work per the project schedule. The separate contractor's delays caused a delay to the critical path of the project and, as a result, the contractor seeks an initial decision from the IDM regarding this claim, which includes an equitable adjustment in the contract time of XX days and an increase in the contract sum of $XX for delay damages.

  B) **Rule**

  Section 3.1 of the A101 indicates that the date of commencement of the work is the owner's notice to proceed to the contractor. Section 3.3 of the A101 identifies the contractual duration of the contractor's work.

Section 6.1.3 of the A201 requires the contractor to review the construction schedule of the owner's separate contractor to revise the schedule, as necessary, to allow for the coordination of work between the contractor and the separate contractor. The updated schedule shall be the schedule to be used by the contractor, separate contractors, and the owner until subsequently revised. Section 6.2.3 makes the owner liable for costs that the contractor incurs due to a separate contractor's delays, improperly timed activities, damage to the work. or defective construction.

Section 8.3.1 of the A201 requires the architect to decide what time, if any, is owed to the contractor if the contractor is delayed by an act or neglect of the owner or the architect. If contractor disagrees with the architect's position, it may make a claim per Article 15.

Section 15.1.3.1 of the A201 requires the contractor to notice the owner, the architect, and the IDM within 21 days of the occurrence of the event giving rise to the claim or recognition of the event giving rise to the claim. Section 15.2.1 mandates that the contractor shall refer claims to the IDM for a decision. If the contractor disagrees with the initial decision, it may trigger mediation and binding dispute resolution per Article 15 of the A201.

C) **Analysis**

The project schedule, which is attached to the A101 agreement, is a contract document and it requires the owner to complete its demolition of the existing building prior to the contractor's mobilization to the project. The owner issued the contractor a notice to proceed for the project so, per the project schedule, the contractor mobilized to the project within five days hence. Note that Section 3.1 of the A101 sets the contractor's commencement of work to be the date of the owner's notice to proceed. Upon mobilization, the owner had not yet completed the demolition work via its separate contractor. As a result, the contractor could not start its critical foundation work for the project per the project schedule.

The delay section of this claim includes a forensic schedule analysis that proves that the contractor is entitled to a XX calendar day time extension for this critical impact. The delay section of this report calculates the contractor's delay damages associated with this extended duration. The contractor properly coordinated with the owner in the development of the project schedule, as required per Section 6.1.3 of the A201, to confirm that the owner's demolition had to be complete upon the contractor's mobilization to the site. The owner failed to meet this obligation. Per Section 6.2.3, the owner is liable for the costs that the contractor incurs due to a separate contractor's delays.

The contractor made the architect aware of this delay in the first OAC meeting at the project but the architect has yet to extend the contract time on the project, even though, per Section 8.3.1 of the A201, the architect shall provide the contractor with such relief if the contractor is delayed by an act or neglect of the owner or the architect.

The contractor previously noticed the owner, the architect, and the IDM of a claim for this issue in accordance with Section 15.1.3.1 of the A201. A copy of this notice is attached herein. Per Section 15.2.1 of the A201, the contractor requests the IDM render an initial decision of this claim.

D) **Conclusion**

Per Section 15.2.1 of the A201, the contractor hereby requests an initial decision from the IDM on this claim related to delay caused by the owner's separate demolition contractor. This delay caused a XX calendar day critical path impact to the project schedule. The contractor requests a modification of the contract time of this number of days and an increase in the contract sum of $XX for related delay damages. If the IDM rejects this claim, the contractor will trigger mediation and, if mediation does not result in settlement, arbitration per the terms of Article 15 of the A201.

## D. Force Majeure Issues

Force majeure issues result from impacts that are beyond the control of both the contractor and the owner, such as abnormal weather, governmental shutdowns, strikes, pandemics, etc. Proving force majeure impacts generally involves establishing that: (1) a force majeure impact did in fact take place; (2) the impact affected the contractor's work; and (3) the owner-contractor agreement allows the contractor to recover time under most contracts and money under some contracts. Establishing such impacts is often done through published information from reliable outside sources that confirm the force majeure issue took place, such as confirmation of abnormal weather from the National Oceanic and Atmospheric Administration (NOAA) or Weather Underground.

**Hypothetical Force Majeure Issue: Abnormal Weather:** The owner retains the contractor under a design-bid-build A102 agreement (design-bid-build with a GMP) that incorporates the A201 general conditions. The project involves the construction of high-rise mixed-use project that has a post-tensioned concrete frame. During the construction phase of work, the contractor's concrete work is slowed by abnormally wet weather during the months of April and May.

**Sample Entitlement Outline:**

A) **Conclusion**

The contractor seeks an initial decision from the IDM for impacts that the contractor has sustained related to abnormal weather that impacted the concrete work at the project during the months of April and May. This abnormally wet weather, which constitutes a force majeure, caused a XX time delay to the critical path of the project, as demonstrated herein.

B) **Rule**

Per Section 8.3.1 of the A201, if the contractor is delayed at any time in the commencement or progress of the work by adverse weather conditions documented in accordance with Section 15.1.6.2, the architect shall extend the contract time as the architect may determine.

Section 8.3.2 of the A201 notes that claims relating to time shall be made in accordance with applicable provisions of Article 15. Section 15.1.6.2 of the A201 states that if adverse weather conditions are the basis for a claim for additional time, such claim shall be documented by data substantiating that (1) the weather conditions were abnormal for the period of time; (2) such conditions could not have been reasonably anticipated; and (3) the abnormal weather had an adverse effect on the scheduled construction. Section 15.1.3.1 of the A201 requires the contractor to notice the owner, the architect, and the IDM within 21 days of the occurrence of the event giving rise to the claim or recognition of the event giving rise to the claim. Section 15.2.1 mandates that the contractor shall refer claims to the initial decision-maker for a decision. If the contractor disagrees with the initial decision, it may trigger mediation and binding dispute resolution per Article 15 of the A201.

C) **Analysis**

The project is located in Dallas, Texas. Per NOAA and Weather Underground, the average precipitation for April and May is 2 inches and 2.5 inches, respectively. Based on these normal weather conditions, the contractor appropriately anticipated three days of weather delays for these two months. The empirical data from both sources is attached herein.

Section 15.1.6.2 of the A201 includes a three-part element test for proving weather delays. The first element is to establish that the weather was indeed abnormal. The actual precipitation for April and May was 3.2 inches and 6.1 inches, respectively; thus, the weather was indeed abnormal when compared to the averages of 2 inches and 2.5 inches. Because of these abnormal conditions, the contractor's critical concrete work was delayed by a total of XX workdays, or XX unanticipated workdays. When you convert this amount of work days into calendar days, it yields XX calendar days (XX workdays / 5 workdays per week x 7 calendar days per week). The second element is foreseeability. The abnormal weather in April and May was unforeseeable. Reasonable contractors contemplate average weather conditions when bidding commercial projects of this nature, and that is what the AIA contract forms consider. The third element considers whether the abnormal weather actually affected the contractor's work. For instance, if there is abnormal weather but it had no effect on the critical work activities at the site, there is no impact. However, the opposite is true here as the contractor was unable to work on the critical concrete work during the abnormal weather days. The contractor's daily reports, which are attached herein, confirm that contractor's concrete subcontractor was rained out for a total of XX workdays between April and May, which confirms that the abnormal weather had a XX workday, or XX calendar day impact on construction. Please refer to the delay section of this report for a complete forensic schedule analysis. Note that the contractor is not seeking delay damages at this time so there is no damages section of the claim.

The contractor made the architect aware of this delay during the OAC meetings during April and May, but the architect has yet to extend the contract time on the project, even though, per Section 8.3.1 of the A201, the architect shall provide the contractor with such relief as the architect so determines if the contractor is delayed by abnormal weather.

The contractor previously noticed the owner, the architect, and the IDM of a claim for this issue in accordance with Section 15.1.3.1 of the A201. A copy of this notice is attached herein. Per Section 15.2.1 of the A201, the contractor requests the IDM render an initial decision of this claim.

D) **Conclusion**

Per Section 15.2.1 of the A201, the contractor hereby requests an initial decision from the IDM on this claim related to abnormal weather delays in the months of April and May, which impacted the substantial completion date by XX calendar days. This claim is supported by weather data from reliable sources in accordance with 15.1.6.2. If the IDM rejects this claim, the contractor will trigger mediation and, if mediation does not result in settlement, arbitration per the terms of Article 15 of the A201.

**Hypothetical Force Majeure Issue: Unavoidable Casualty:**     The owner retains the contractor under a design-bid-build A101 agreement that incorporates the A201 general conditions. The project involves the construction of a multi-family apartment complex. Approximately two months before the contractor was scheduled to achieve substantial completion a tornado event impacted the project site and caused damage to the roofing, siding, windows, and landscaping, which caused a XX calendar day delay to the substantial completion date. The contractor recovered the costs associated with the damage through the owner's property insurance policy under Section 11.5 of the A201 general conditions, so the only disputed issue is the reasonable extension of time related to this event.

**Sample Entitlement Outline:**

A) **Conclusion**

The contractor seeks an initial decision from the IDM for time-related impacts that the contractor has sustained related to the recent tornado that damaged significant work at the project. This tornado caused a XX calendar day delay to the critical path of the project, as demonstrated herein. Note that the contractor recovered the costs associated with the damages from the property insurance carrier, so this claim is limited to a proposed adjustment to the contract time.

B) **Rule**

Per Section 8.3.1 of the A201, if the contractor is delayed at any time in the commencement or progress of the work by an unavoidable casualty, the architect shall extend the contract time as the architect may determine.

Section 8.3.2 of the A201 notes that claims relating to time shall be made in accordance with applicable provisions of Article 15. Section 15.1.3.1 of the A201 requires the contractor to notice the owner, the architect, and the IDM within 21 days of the occurrence of the event giving rise to the claim or recognition of the event giving rise to the claim. Section 15.2.1 mandates that the contractor shall refer claims to the IDM for a decision. If the contractor disagrees with the initial decision, it may trigger mediation and binding dispute resolution per Article 15 of the A201.

C) **Analysis**

The project is located in the north of Denver, Colorado. This past June a tornado passed through the project site and caused significant damage to the work at the project. Attached are NOAA and Weather Underground articles that verify the time, date, and path of the tornado. Also attached are the contractor's daily reports through the delay period and daily photographs of the damaged site.

The delay section of this claim reviews the critical path impact of this weather event through a detailed forensic schedule analysis. In addition, the contractor's daily reports are attached that confirm that no work took place on the project for XX working days, or XX calendar days. The contractor made the architect aware of this delay immediately after this event but the architect has yet to extend the contract time per Section 8.3.1 of the A201, which requires the architect to extend the contract time as the architect so determines for unavoidable casualties, such as the subject tornado.

The contractor previously noticed the owner, the architect, and the IDM of a claim for this issue in accordance with Section 15.1.3.1 of the A201. A copy of this notice is attached herein. Per Section 15.2.1 of the A201, the contractor requests the IDM render an initial decision of this claim.

D) **Conclusion**

Per Section 15.2.1 of the A201, the contractor hereby requests an initial decision from the IDM on this claim related to the recent tornado that passed over the construction site, which impacted the substantial completion date by XX calendar days. This claim is supported by weather data from reliable sources as well as the contractor's daily reports and photographs. If the IDM rejects this claim, the contractor will trigger mediation and, if mediation does not result in settlement, arbitration per the terms of Article 15 of the A201.

## E. Change Order Negotiation Issues

Nearly all construction projects involve additive or deductive change orders between the owner and contractor. When the parties agree that a change order is owing, proving entitlement is straightforward, as a change order represents recognition of a cost and/or time impact to the contractor's work. Disputes often arise regarding the magnitude of the impact to the contractor's work, which can bring the change order negotiation to a stalemate. It is critical for contractors to understand what remedies it has regarding change order work in order to mitigate cash flow constraints caused by such a directive.

**Hypothetical Owner Performance Issue: Failed Change Order Negotiation:** The owner retains the contractor under an AIA A102 agreement (design-bid-build with a GMP) for an industrial warehouse project. The owner, through its architect, sought to add additional mechanical work to the contractor's scope of work through the traditional negotiation process, but the parties could not reach an agreement to the estimated impact to the contract sum and contract time. As a consequence, the architect issued the contractor a Construction Change Directive (CCD) for the mechanical work, so the contractor completed the work. Thereafter, the contractor provided the owner and the architect with backup related to the work, but the parties could still not reach an agreement on the cost and time impacts so the architect issued an interim agreement and the owner paid the contractor accordingly. Because the contractor disagreed with the architect's interim agreement position, it triggered dispute resolution.

**Sample Entitlement Outline:**

A) **Conclusion**

The contractor seeks an initial decision from the IDM related to the cost and time impacts associated with the mechanical work CCD that the owner and architect issued to the contractor several months ago. The requested amounts are above and beyond what the architect approved in its interim decision.

B) **Rule**

Per Section 7.3.2 of the A201, the architect and the owner may issue the contractor a CCD in the absence of total agreement on the terms of a change order. Section 7.3.4 notes that if the contractor disagrees with the method for adjustment in the contract sum, the architect shall determine the adjustment. If the contractor disagrees with the architect's position, Section 7.3.5 allows the contractor to make a claim in accordance with Article 15 requirements. Section 7.3.9 further allows the contractor to make an Article 15 claim if it disagrees with the architect's interim decision.

Section 15.1.3.1 of the A201 requires the contractor to notice the owner, the architect, and the IDM within 21 days of the occurrence of the event giving rise to the claim or recognition of the event giving rise to the claim. Section 15.2.1 mandates that the contractor shall refer claims to the IDM for a decision. If the contractor disagrees with the initial decision, it may trigger mediation and binding dispute resolution per Article 15 of the A201.

C) **Analysis**

Because the owner and the contractor could not agree to the terms of the additive mechanical work modification, the owner and the architect issued the contractor a CCD, as allowed under Section 7.3.2 of the A201. Immediately thereafter, the contractor proceeded with the work and presented all cost and time impacts to the owner and the architect for review.

Upon review of the submitted information, the architect disagreed with the contractor's position in terms of the cost and time impacts related to the work so it rendered an interim decision in favor of the owner. The contractor disagrees with this interim decision and seeks relief under Section 7.3.9 and Article 15.

The delay section of this claim includes a detailed forensic schedule analysis to prove that the contractor is owed an extension of contract time of XX days. The damages section of this claim includes actual cost backup for the mechanical improvements and the contractor's extended general conditions related to the delay period. When the contractor's time and cost impacts are compared with the architect's interim decision, it results in an additional extension of the contract time of XX days and an additional increase to the contract sum of $XX.

The contractor previously noticed the owner, the architect, and the IDM of a claim for this issue in accordance with Section 15.1.3.1 of the A201. A copy of this notice is attached herein. Per Section 15.2.1 of the A201, the contractor requests the IDM render an initial decision of this claim.

D) **Conclusion**

Per Section 15.2.1 of the A201, the contractor requests an initial decision from the IDM on this claim related to the time and cost impacts of the mechanical work CCD. Based on the information presented herein, the contractor is entitled to an addition extension in the contractor time of XX calendar days and an increase to the contract sum of $XX. If the IDM rejects this claim, the contractor will trigger mediation and, if mediation does not result in settlement, then arbitration per the terms of Article 15 of the A201.

## III. Typical Owner Claims Against Contractors

The most common type of claim that owners assert against contractors relate to quality issues, schedule issues, administrative issues, design issues, third-party issues, and change order negotiation issues. The following discussion assumes that the claimant is the owner, and the owner-contractor agreement is an AIA document that incorporates the A201 general conditions.

### A. Quality Issues

Construction contracts require the contractor to install work in accordance with the contract documents. When contractors fail to meet this requirement, owner remedies include the stoppage of the work, withholding earned funds, or correction of the default after proper notice. To prove entitlement for a quality claim, the owner must compare the installed work with the requirements of the contract documents or approved submittals and clearly illustrate the difference between the two perhaps change to manufacturer's recommendations or shop drawings; not a huge deal. Quality issues can be subjective in nature and interpretations can vary, so the retention of objective third-party experts can be helpful to support owner contentions.

**Hypothetical Contractor Quality Issue: Lack of Open Head Joint Weeps in the Brick Veneer:** The owner retained the contractor under an AIA A101 agreement (design-bid-build for a stipulated sum) to construct a four-story office building that is clad with full brick veneer. Shortly after the contractor commences with the brick veneer work, the architect notices that the contractor has failed to install open head joint weeps above all through wall flashing in the brick veneer, which is clearly detailed in the architectural plans. Per the architectural plans, open head joint weeps are required every 24 inches on center. Accordingly, the architect rejects the work and notices the contractor of various remedies that the owner maintains under the contract for defect work issues.

**Sample Entitlement Outline:**

A) **Conclusion**

   The contractor failed to install open head joint weeps every 24 inches on center above through wall flashing throughout the exterior brick veneer installed to date. The architect rejects this work and intends to withhold funds from future contractor payment applications to cover the estimated cost to correct this issue. The architect reminds the contractor that the owner has the right under the A201 to stop the masonry work and arrange for the correction of the defects if the contractor fails to take prompt action.

B) **Rule**

   Section 3.1.2 of the A201 requires the contractor to perform the work in accordance with the contract documents. Section 12.2.1 of the A201 mandates that the contractor shall promptly correct work rejected by the architect for failing to conform to the requirements of the contract documents at the contractor's own expense. Sections 9.4.1 and 9.5.1 of the A201 allow the architect to withhold certification of the contractor's payment application in whole or in part because of, among other items, defective work not remedied. In addition, Section 2.4 of the A201 allows the owner to stop the work related to the defect upon written direction from the owner to the contractor if the contractor fails to correct the rejected work per Section 12.2. The owner may keep a stop work order in place until the contractor addresses the cause for the defective work.

Also, Section 2.5 of the A201 permits the owner to carry out the corrective work if the owner issues the contractor a ten-day cure notice and the contractor fails to commence and continue correction of the default within this time period. Note that this section requires the owner to get the architect's approval before undertaking such action. The architect must also approve the amount the owner wishes to backcharge the contractor once it completes the correction.

C) **Analysis**

The contract documents for the project, including the architectural drawings and masonry specifications, prescribe the installation of open head joint weeps every 24 inches on center above through wall flashing incorporated into the exterior brick veneer. Screenshots of these requirements are shown below.

As evidenced in the photographs appended herein, the contractor has failed to install these open head joint weeps above any through wall flashing incorporated into the brick veneer to date. As a result of this faulty installation, any moisture that gets into the wall cavity is trapped and may find its way into the interior of the building. Section 3.1.2 of the A201 requires the contractor to install its work per the contract documents. Because the contractor's masonry work does not conform with the contract requirements, the architect hereby rejects this work.

Per Section 12.2.1, the contractor shall promptly correct work rejected by the architect for failing to conform to the requirements of the contract documents at the contractor's own expense. Costs of correcting such rejected work, including additional testing and inspections, the cost of uncovering and replacement, and compensation for the architect's services and expenses made necessary thereby, shall be at the contractor's expense.

Until this defect is corrected, the architect intends to withhold funds against the contractor's future payment applications, as allowed per Sections 9.4.1 and 9.5.1 of the A201. The amount of this withholding will represent the estimated cost to cut in the open head joint weeps into the brick veneer at the required intervals. See the cost section of this report for a detailed breakdown of this estimate.

In the event that the contractor fails to promptly correct the defective work and install proper weeps on the remaining work, the architect reminds the contractor that the owner has the right to halt the masonry work per Section 2.4 of the A201 upon the owner's written direction to the contractor. A stoppage of this sort could remain in effect until the contractor corrects the cause of the defective installations.

In addition, the owner has the right under Section 2.5 to carry out the corrective work if the owner issues the contractor a ten-day cure notice to commence and continue correction of the rejected masonry work and the contractor takes no substantive action. The owner is then entitled to backcharge the costs associated with the remedial effort to the contractor. Note that the owner's Section 2.5 remedies are subject to the architect's approval.

D) **Conclusion**

In sum, the architect rejects the contractor's masonry work installed to date for failure to install open head joint weeps above all through wall flashings throughout the masonry veneer installed to date. Per Section 12.2.1 of the A201, the contractor shall promptly commence correction of this work. Until this defective work is remedied, the architect intends to withhold funds from the contractor's future payment applications pursuant to Sections 9.4.1 and 9.5.1 of the A201 for the estimated cost to correct the non-compliant work. Lastly, in the event that the contractor fails to address this defective work in a timely manner, the owner may halt the masonry work if the defective installations continue and it may take over the corrective effort per Sections 2.4 and 2.5 of the A201.

**Hypothetical Contractor Quality Issue: Floor Flatness and Floor Levelness Issues at Elevated Concrete Slabs:** The owner retained the contractor under an AIA A134 agreement (cost-plus CM agreement) that incorporates the AIA A201 general conditions to construct a

15-story residential condominium building. After the contractor placed the first two floor slabs and it issued its floor flatness and floor levelness reports, the architect determined that the contractor was not achieving average or even minimum tolerance requirements on a majority of the slab area, which were industry-accepted floor flatness and floor levelness requirements for concrete floors that were to receive carpet.

**Sample Entitlement Outline:**

A) **Conclusion**

The architect reviewed the contractor's floor flatness and floor levelness reports related to Levels 2 and 3 on the project and it is clear that the majority of the slab area fails to meet the average or even minimum floor flatness and floor levelness requirements required per specification. Accordingly, the architect rejects this work and will withhold funds from the next contractor payment application to cover the estimated cost to correct this issue. The architect reminds the contractor that the owner has the right under the A201 to stop the concrete work and arrange for the correction of the defects if the contractor fails to take prompt action.

B) **Rule**

Section 3.1.2 of the A201 requires the contractor to perform the work in accordance with the contract documents. Section 12.2.1 of the A201 mandates that the contractor shall promptly correct work rejected by the architect for failing to conform to the requirements of the contract documents at the contractor's own expense. Sections 9.4.1 and 9.5.1 of the A201 allow the architect to withhold certification of the contractor's payment application in whole or in part because of, among other items, defective work not remedied. In addition, Section 2.4 of the A201 allows the owner to stop the work, upon written direction from the owner to the contractor, if the contractor fails to correct rejected work per Section 12.2 or if the contractor repeatedly fails to carry out the work per the contract documents. The owner can keep the stop work order in effect until the contractor addresses the cause for the defective work.

Also, Section 2.5 of the A201 permits the owner to carry out the corrective work if the owner issues the contractor a ten-day cure notice regarding the defective work and the contractor fails to commence and continue correction of the default within this period. Note that this section requires the owner to get the architect's approval before it undertakes such action. The architect must also approve the amount the owner wishes to backcharge the contractor once it completes the corrective effort.

C) **Analysis**

The contract documents for the project, including the structural drawings and the concrete specifications, require the contractor to meet certain concrete floor flatness and floor levelness requirements, and require the contractor to submit reports to the owner and architect of floor flatness and floor levelness readings shortly after concrete placement. These concrete requirements are shown below.

Based on my review of the contractor's third party reports for the elevated slabs at Level 2 and 3, the vast majority of surface area is not within tolerance. This non-compliant work is further supported by the attached photographs, which clearly show patent waviness in the concrete work. Because the contractor's concrete work does not conform with the contract requirements, the architect hereby rejects this work for being defective.

Per Section 12.2.1, the contractor shall promptly remediate the non-compliant surfaces of the Levels 2 and 3 slabs at the contractor's own expense. Costs of correcting such rejected work (through means such as grinding high areas and the application of an acceptable self-leveling compound for low areas), including additional testing and inspections, the cost of repair, and compensation for the architect's services and expenses made necessary thereby, shall be at the contractor's expense per Section 12.2.1 of the A201.

Until this defective work is corrected, the architect intends to withhold funds against the contractor's payment applications, as allowed per Sections 9.4.1 and 9.5.1 of the A201. The amount of this withholding will represent the estimated cost to patch and grind the defective surface to bring it into conformance with the contract documents. See the cost section of this report for a detailed breakdown of this estimate. The withholding will cease once the contractor completes this work.

In the event that the contractor fails to promptly correct the defective work and place compliant work on the remaining floors, the architect reminds the contractor that the owner has the right to halt the concrete work upon the owner's written direction, per Section 2.4 of the A201. A stoppage of this sort could remain in effect until the contractor corrects the cause of the defective placements.

In addition, the owner has the right under Section 2.5 to carry out the corrective work if the owner issues the contractor a ten-day cure notice to commence and continue correction of the rejected concrete surfaces and the contractor takes no substantive action. The owner is then entitled to backcharge the costs associated with the owner's remedial effort to the contractor. Note that the owner's Section 2.5 remedies are subject to the architect's approval.

D) **Conclusion**

The architect rejects the non-conforming concrete surface area on Levels 2 and 3 for failure to conform to the contract documents relating to the floor flatness and floor levelness criteria. Per Section 12.2.1 of the A201, the contractor shall promptly commence correction of this work. Until this defective work is remedied, the architect intends to withhold funds from the contractor's payment applications pursuant to Sections 9.4.1 and 9.5.1 of the A201 for the estimated cost to correct the non-compliant work. Lastly, in the event that the contractor fails to address this defective work in a timely manner, the owner may halt the elevated concrete work if the defective installations continue and it may also take over the corrective effort per Sections 2.4 and 2.5 of the A201.

## B. Schedule Issues

Construction contracts typically require the contractor to achieve substantial completion of the project within a specified duration and in substantial accordance with an approved project schedule. When a project is delayed for contractor-caused reasons, owners often assert delay claims against the contractor for liquidated damages or actual delay damages, depending on the terms of the contract. In order to prove a contractor-caused delay, the owner can compare the requirements of the baseline schedule (with excusable time extensions) or recent schedule updates with the as-built construction progress to identify contractor-caused slippage to the substantial completion date of the subject project.

**Hypothetical Contractor Quality Issue: Project Delays (Liquidated Damages):** The owner retained the contractor under an AIA A101 agreement (design-bid-build with a stipulated sum) that incorporates the AIA A201 general conditions to construct a 12-story office building that has a steel frame superstructure. The A101 agreement stipulates a liquidated damage rate of $10,000 per calendar day and a construction duration of 24 months. Approximately six months into the project, the contractor's schedule started to slip because its steel subcontractor was unable to mobilize sufficient workers to the project for no purported reason. As of the 24-month mark, the contractor was six months behind schedule. Note that liquidated damages are dealt with by the architect via the contractor's payment application process under certain AIA contracts.

**Sample Entitlement Outline:**

A) **Conclusion**

The architect certifies the contractor's current payment application of $400,000 for $150,000 and withholds $250,000 as a liquidated damage assessment as reasonable evidence exists that the contractor will complete the project six months late due to contractor-caused delays. Moreover, if the contractor fails to recover this delay, all future payment applications will be withheld in full as the contractor's unpaid contract balance of $1,800,000 matches a six-month liquidated damage assessment for which the contractor is responsible unless it can recover the schedule.

B) **Rule**

Section 3.3.1 of the A101 notes that the contractor shall substantially complete the project within 24 months. Section 4.5 of the A101 lists the liquidated damages rate for the project at $10,000 per calendar day for every day that the contractor fails to achieve substantial completion for contractor-caused reasons.

Per Section 9.4.1, the architect shall respond to the contractor within seven days of receipt and either: (1) certify the payment application in full; (2) certify the payment application in part and advise the contractor and owner of the reasons for Section 9.5.1 withholdings; or (3) withhold certification of the entire payment application amount and advise the contractor and owner of the reasons for Section 9.5.1 withholdings.

Section 9.5.1 allows the architect to withhold a certificate for payment in whole or in part to protect the owner because of, among other items, reasonable evidence exists that the contractor will not complete the work within the contract time and the contractor's unpaid balance would not be adequate to cover actual or liquidated damages for the anticipated delay.

C) **Analysis**

Section 3.3.1 of the A101 agreement between the owner and contractor requires the contractor to achieve substantial completion of the project within 24 months. The contractor's most recent schedule update, which is attached herein, reflects a six-month delay to the substantial completion date. This delay is a result of the contractor's slow production on the structural steel work, which took six months longer than scheduled, which was discussed in numerous OAC meetings. Note the contractor has no pending claims for an extension to the contract time.

The contractor issued its current payment application to the architect three days ago in the amount of $400,000. Per Section 9.4.1, the architect certifies for $150,000 and plans to withhold $250,000 as reasonable evidence exists that the contractor will not complete the work within the contract time and the contractor's unpaid balance is not adequate to cover liquidated damages for the anticipated delay.

As noted above, the project is currently six months late. Section 4.5 of the A101 lists the liquidated damage rate of $10,000 per calendar day. When six months, or 180 calendar days, is multiplied by $10,000, it equates to an assessment of $1,800,000. Based on the contractor's last approved payment application, which is attached herein, the contractor's unpaid contract balance on the project is approximately $1,950,000, and this figure is inclusive of earned retainage. Thus, when $1,800,000 is deducted from $1,950,000, the balance is $150,000. Accordingly, the architect's certification of $150,000 of the contractor's current payment application is because $1,800,000 must be withheld to protect the owner. Note that all future payment applications will be withheld, unless the contractor is able to partially recover the delay, in which case the withholding will be reduced by $10,000 for each calendar day of compression.

D) **Conclusion**

The architect certifies the contractor's current payment application for $150,000 and withholds $250,000 due to a liquidated damage assessment because the contractor's most recent schedule update shows the contractor completing six months late. Thus, the contractor's remaining unpaid contract balance of $1,800,000 will be withheld unless the contractor is able to partially recover the schedule, in which case the withholding will be reduced by $10,000 for each day of recovery.

**Hypothetical Contractor Quality Issue: Project Delays (Actual Damages):** The owner retained the contractor under an AIA A101 agreement (design-bid-build with a stipulated sum) that incorporates the AIA A201 general conditions to construct a 30-story office building that has a steel frame superstructure. The A101 agreement does not stipulate a liquidated damage rate. The contractor's schedule started to slip because the steel subcontractor was unable to mobilize sufficient workers to the project for no purported reason. Per the A101, the contractor's contract time to achieve substantial completion was 36 months. The contractor achieved substantial completion at month 38, two months late for contractor-caused reasons.

**Sample Entitlement Outline:**

A) **Conclusion**

The architect certifies the contractor's final payment application for $1,200,000, and withholds $150,000 for the actual damages that the owner incurred as a result of the contractor's two-month delay on the project.

B) **Rule**

Section 3.3.1 of the A101 notes that the contractor shall substantially complete the project within 36 months. Section 4.5 of the A101 does not stipulate a liquidated damages rate for the project. Per Section 9.4.1, the architect shall respond to the contractor within seven days of receipt and either: (1) certify the payment application in full; (2) certify the payment application in part and advise the contractor and owner of the reasons for Section 9.5.1 withholdings; or (3) withhold certification of the entire payment application amount and advise the contractor and owner of the reasons for Section 9.5.1 withholdings.

Section 9.5.1 allows the architect to withhold a certificate for payment in whole or in part to protect the owner because of, among other items, reasonable evidence exists that the contractor will not complete the work within the contract time and the contractor's unpaid balance would not be adequate to cover actual or liquidated damages for the anticipated delay.

C) **Analysis**

Section 3.3.1 of the A101 agreement between the owner and the contractor requires the contractor to achieve substantial completion of the project within 36 months. The contractor achieved substantial completion in 38 months, or two months late. This delay is a result of the contractor's slow steel work. The contractor has no pending claims for an extension of the contract time.

The contractor issued the architect its final payment application yesterday in the amount of $1,350,000. Per Section 9.4.1, the architect certifies the application for $1,200,000 and withholds $150,000 for actual damages that the owner incurred because of the contractor's late completion. Section 9.5.1 allows the architect to withhold contract funds to cover actual damages incurred by the owner for the contractor's late completion. The owner's damages include the costs related to keeping the owner's representative and architect on site for two additional months. Invoices for these charges that amount to $150,000 and confirmation of payment are attached herein.

D) **Conclusion**

The architect certifies the contractor's final payment application in the amount of $1,200,000 and withholds $150,000 for actual damages that the owner incurred as a result of the contractor's late completion of work.

## C. Administrative Issues

Owners often withhold earned funds from contractors due to contractor administrative issues such as: the failure to make timely payment to subcontractors or vendors for work performed and previously paid for; failure to properly clean-up the site; or failure to properly administrate the submittal process. Proof of entitlement for this type of claim can be done by detailing the contract requirement that outlines the contractor's administrative duty and then providing evidence that confirms the contractor breached this duty, such as documentation of claims by subcontractors or suppliers, photographs of an unkempt jobsite, or copies of submittals that fail to adhere to the contract requirements.

**Hypothetical Contractor Administrative Issue: Failure to Make Payment:** The owner retained the contractor under an AIA A102 agreement (design-bid-build with a GMP) that incorporates the AIA A201 general conditions to construct a retail center. When the project was approximately 20% complete, two subcontractors formally noticed the owner of the contractor's failure to make payment to them for work related to the last funded payment application. Also, the architect received verbal notification that one of the contractor's vendors had not yet been paid, as required by contract.

**Sample Entitlement Outline:**

A) **Conclusion**

The architect certifies the contractor's current payment application for $85,000 and withholds $275,000 due to subcontractor and vendor claims for monies owed relating to previously funded payment applications.

B) **Rule**

Per Section 9.3.3 of the A201, the contractor warrants that, upon a submission of an application for payment, to the best of the contractor's knowledge, that all work for which has been certified and paid for by the owner is free and clear of liens, claims, security interests, or encumbrances.

Per Section 9.4.1, the architect shall respond to the contractor within seven days of receipt and either: (1) certify the payment application in full; (2) certify the payment application in part and advise the contractor and owner of the reasons for Section 9.5.1 withholdings; or (3) withhold certification of the entire payment application amount and advise the contractor and the owner of the reasons for Section 9.5.1 withholdings.

Section 9.5.1 allows the architect to withhold a certificate for payment in whole or in part to protect the owner because of, among other items, the failure of the contractor to make payments properly to subcontractors or suppliers for labor, materials, or equipment.

Section 9.5.3 of the A201 notes that when the reasons for withholding certification are removed, certification will be made for amounts previously withheld. Also, per Section 9.5.4, if the architect withholds certification under Section 9.5.1, the owner may issue joint checks to the contractor and to any subcontractor or supplier to whom the contractor failed to make payment for work properly performed or material or equipment suitably delivered. If the owner makes payments by joint check, the owner shall notify the architect and the contractor shall reflect such payment on its next application for payment.

C) **Analysis**

The contractor's excavation and concrete subcontractors provided detailed written notification to the owner and the architect that the contractor had failed to make payment to them for the last owner-funded payment application. A copy of these notifications is found herein. In addition, the rebar vendor retained directly by the contractor verbally told the architect that it has yet to receive payment. The two formal claims amount to $225,000 and the informal claim amounts to $50,000.

The architect hereby exercises its right under Section 9.4.1 of the A201 to certify part of the contractor's application for payment and to withhold the balance for reasons allowed under Section 9.5.1. Per Section 9.5.1.3 of the A201, the architect can withhold funds in the event that the contractor fails to properly pay subcontractors or suppliers for work previously completed, certified, and funded by the owner. Accordingly, the architect withholds $275,000 from the contractor for claims made by the contractor's concrete subcontractor, excavation subcontractor, and rebar vendor.

Per Section 9.5.3 of the A201, once the contractor provides proof of payment to these parties, certification will be made for this withholding. Note that the owner has the right to issue joint checks to these parties in the event that proper payment has not been made by the contractor.

D) **Conclusion**

In sum, the architect certifies the contractor's current payment application for $85,000 and withholds $275,000 due to subcontractor and vendor claims for monies owed relating to previously paid applications for payment. Please investigate these claims immediately as it constitutes maladministration of the contract if the claims are valid.

**Hypothetical Contractor Administrative Issue: Failure to Clean Up:** The owner retained the contractor under an AIA A133 agreement (CM at risk with a GMP) that incorporates the AIA A201 general conditions to construct a data center. During the course of construction, the architect and the owner continually advised the contractor of its failure to keep the work area free from accumulation of waste materials and rubbish. Eventually, the owner decided to arrange for a vendor to clean up the project.

**Sample Entitlement Outline:**

A) **Conclusion**

The contractor has continually failed to maintain work areas that are free of accumulation of waste materials and rubbish. As a result, the owner arranged for a vendor to clean the site and the owner expects reimbursement of $XX for this cost, as stipulated per contract.

B) **Rule**

Section 3.15.1 requires the contractor to keep the work area free from accumulation of waste materials and rubbish caused by operations under the contract. Section 10.2.1 of the A201 mandates that the contractor take reasonable precautions for the safety of employees and other persons involved with the project.

Per Section 3.15.2 of the A201, if the contractor fails to clean up as provided in the contract documents, the owner may do so and the owner shall be entitled to reimbursement from the contractor. Section 6.3 of the A201 further permits the owner to clean up the work and if a dispute arises among the contractor, separate contractors, and the owner as to the responsibility under their respective contracts for maintaining the premises and surrounding area free from waste materials and rubbish, the architect will allocate the cost among those responsible.

Per Section 9.4.1, the architect shall respond to the contractor within seven days of receipt of its payment application and either: (1) certify the payment application in full; (2) certify the payment application in part and advise the contractor and the owner of the reasons for Section 9.5.1 withholdings; or (3) withhold certification of the entire payment application amount and advise the contractor and the owner of the reasons for Section 9.5.1 withholdings.

Section 9.5.1 allows the architect to withhold a certificate for payment in whole or in part to protect the owner because of, among other items, damage to the owner caused by the contractor.

C) **Analysis**

The contractor has failed to maintain its work area free from waste materials and rubbish, as is required by Section 3.1.15 of the A201. The contractor's breach has placed the workers on the project in danger, which the contractor has a duty to prevent under Section 10.2.1 of the A201. The owner has alerted the contractor many times during OAC meetings of this concern, as shown in the attached meeting minutes. Also attached are photographs taken early this week of the project site that confirms that the contractor failed to take heed of these recommendations as waste and debris fill the majority of areas on the project.

As a result of the contractor's neglect, the owner retained a separate vendor to clean up the rubbish and waste over this past weekend and per Section 3.15.2 of the A201, the owner seeks reimbursement from the contractor for this cost. Attached is the invoice from the cleaning contractor in the amount of $XX that the owner has funded. A copy of the owner's check issued to the cleaning vendor is included herein as well. If the contractor fails to reimburse the owner, it will present the cost to the architect and the architect shall determine if the owner has damages due to the neglect of the contractor and it will consider an offset to the contractor's future payment applications per Sections 9.4.1 and 9.5.1 of the A201.

D) **Conclusion**

The contractor failed to maintain a clean and safe worksite so the owner exercised its right to clean the project. The owner now seeks reimbursement of $XX for these costs per Section 3.1.15 of the A201.

## D.  Contractor Design Issues

Design-build contractors take on full design responsibilities for construction projects, which is based on design criteria provided by project owners. Design-bid-build contractors and construction managers often take on delegated design responsibilities for discrete items per the contract documents such as the engineering of fire suppression systems, shoring, metal framing, precast, etc. If a contractor's design does not adhere to the design criteria or performance requirements outlined in the contract documents, the owner can assert a design error or omission claim against the contractor. To prove entitlement for such a claim, the owner must compare the design criteria per contract against the contractor's design to identify design elements that fall short of the mandated criteria. Or, if the contract documents outline a performance specification, the owner must compare the performance requirements with the actual performance of the system designed and installed by the contractor to identify performance that falls below the stipulated requirements.

**<u>Hypothetical Contractor Design Issue: Design Error on Fire Suppression Plans</u>:** The owner retained the contractor under an AIA A133 agreement (CM at risk with a GMP) that incorporates the AIA A201 general conditions to construct an industrial building. The construction documents delegate the design of the fire suppression system to the contractor. Specifically, the contract drawings note, "Provide a complete wet-pipe automatic sprinkler system throughout the project area. The system shall be designed as per NFPA-13 light hazard." As the project neared substantial completion, the architect determined that the fire suppression system, as installed, failed to meet several requirements of NFPA-13 in terms of main size, water treatment, and foundation penetrations.

**<u>Sample Entitlement Outline:</u>**

A) **Conclusion**

The architect rejects the contractor's fire suppression system for its failure to meet the design criteria outlined in the contract documents. The architect plans to withhold $XX in funds from the contractor's next payment application for this defective work that is not remedied.

B) **Rule**

Section 3.1.2 of the A201 requires the contractor to perform work in accordance with the contract documents. Furthermore, Section 3.12.10.1 of the A201 notes that if the contract documents delegate the design to the contractor, the owner and the architect shall specify all performance and design criteria.

Section 12.2.1 of the A201 requires the contractor to promptly correct work rejected by the architect for failing to conform to the requirements of the contract documents. Costs of correcting such rejected work, including additional testing and inspections, the cost of uncovering and replacement, and compensation for the architect's services and expenses made necessary thereby, shall be at the contractor's expense.

Per Section 9.4.1, the architect may withhold certification of the contractor's payment application in whole or in part as provided for in Section 9.5.1. Section 9.5.1 notes that the architect may withhold a certificate of payment in whole or in part because of defective work not remedied.

In addition, Section 2.4 of the A201 allows the owner to stop the work, upon written direction from the owner to the contractor, if the contractor fails to correct rejected work per Section 12.2 or if the contractor repeatedly fails to carry out the work per the contract documents. The owner can keep the stop work order in effect until the contractor addresses the cause of the defective work.

Also, Section 2.5 of the A201 permits the owner to carry out the corrective work if the owner issues the contractor a ten-day cure notice regarding the defective work and the contractor fails to commence and continue correction of the default within this period. Note that this section requires the owner to get the architect's approval before it undertakes such action. The architect must also approve the amount the owner wishes to backcharge the contractor once it completes the corrective effort.

C) **Analysis**

As shown below, the contract documents note that the fire suppression design for the project is delegated to the contractor, and the criteria for the design follows NFPA-13. NFPA-13 stipulates requirements that relate to main size, water treatment, foundation penetrations, and other design parameters. Details of these requirements are listed in the table below.

The photographs attached to this report confirm that the contractor's work does not meet NFPA-13 light hazard requirements in terms of the minimum diameter of the private services main, the water treatment system, and the size of the foundation penetrations. These design and installation flaws will detrimentally affect the performance of the fire suppression system.

The architect rejects the contractor's fire suppression work and demands the contractor promptly correct this work per Section 12.2.1 of the A201. Until this defective work is corrected, the architect intends to withhold funds against the contractor's payment applications, as allowed per Sections 9.4.1 and 9.5.1 of the A201. The amount of this withholding will be $XX, which represents the estimated cost to bring the work into conformance with the contract documents. See the cost section of this report for a detailed breakdown of this estimate. The withholding will cease once the contractor completes this work and demonstrates the work conforms with the design requirements.

In the event that the contractor fails to promptly correct the defective work, the architect reminds the contractor that the owner has the right to halt the work upon the owner's written direction, per Section 2.4 of the A201. A stoppage of this sort could remain in effect until the contractor corrects the cause of the defective placements.

In addition, the owner has the right under Section 2.5 to carry out the corrective work if the owner issues the contractor a ten-day cure notice to commence and continue correction and the contractor takes no substantive action. The owner is then entitled to backcharge the costs associated with the owner's remedial effort to the contractor. Note that the owner's Section 2.5 remedies are subject to the architect's approval.

D) **Conclusion**

The architect rejects the contractor's fire suppression work for failing to comply with the design criteria outlined in the contract documents. The contractor is required to promptly correct this work per the contract, and the architect intends to withhold payment until the defects are cured.

### E. Contractor Impacts on Owner or Owner's Separate Contractors

Standard contract agreements require the contractor to cooperate with separate contractors retained by the owner that are scheduled to perform work on the project concurrently with the work of the contractor. A typical example of this is when an owner on a building project retains a furniture, fixtures, and equipment contractor that is separate and apart from the prime contractor on the project. This is common on educational projects, office projects, and hospitality projects. If the contractor, owner, and separate contractor agree to an overall schedule and the contractor is bound to that schedule per the terms of the owner-contractor agreement, any detrimental action of the contractor to the separate contractor that causes the owner to pay additional money to the separate contract can ultimately be a contractor responsibility. Standard contracts often permit owners to backcharge contractors per the terms of the contract in these instances in order to recover losses. To establish entitlement for this type of claim, an owner must prove that the contractor's actions violated the terms of the agreement and caused damage to the separate contractor, and in turn the owner made payment of the separate contractor's claim.

**Hypothetical Contractor-Caused Delays to a Separate Contractor:**
The owner retained the contractor under an AIA A133 agreement (CM at risk with a GMP) that incorporates the AIA A201 general conditions to construct a high-rise hotel project. Per the terms of the agreement the owner will retain a separate contractor for all FF&E work. The contractor and the separate contractor agreed to coordinate the work per the contractor's schedule. As it turns out, the contractor delivered many of the hotel rooms to the separate contractor several months late due to its painting subcontractor's slow work, which increased the separate contractor's time on the project. Consequently, the separate contractor issued the owner a claim for two months of extended general conditions, which the owner approved and paid and then backcharged to the contractor. Note that this finish work was not on the critical path of the overall project, which ran through the contractor's exterior civil work. Thus, the delay to the separate contractor did not extend the substantial completion date of the project.

**<u>Sample Entitlement Outline for Impacts to Separate Contractor:</u>**
A) **Conclusion**

Because of the slow progress of the contractor's painting work, the owner paid its separate FF&E contractor its claim for two months of extended general conditions. The owner hereby seeks reimbursement of $XX per the terms of the A201 general condition of the A201 for this damage caused to the owner and the separate contractor.

B) **Rule**

The contract is clear in terms of the owner's right to retain separate contractors on the project. Section 1.1.4 notes that the contractor's work is part of the project and the project might involve the work of the owner and separate contractors. Moreover, the owner reserves the right to perform work with its own forces or through a separate contractor per Section 6.1.1.

Section 6.1.1 states that the contractor shall participate with any separate contractors and the owner in reviewing their construction schedules. The contractor shall make any revisions to its construction schedule deemed necessary after a joint review and mutual agreement. The construction schedules shall then constitute the schedules to be used by the contractor, separate contractors, and the owner until subsequently revised.

Section 6.2.3 of the A201 notes that the contractor shall reimburse the owner for costs the owner incurs that are payable to a separate contractor because of the contractor's delays, improperly timed activities, or defective construction. In addition, Section 9.5.1 allows the architect to withhold a contractor's certificate of payment in whole or in part because of damage to the owner or a separate contractor.

In the event that the contractor fails to reimburse this amount to the owner, the owner may recommend that the architect withhold this amount from a future contractor payment application per Sections 9.4.1 and 9.5.1 of the A201.

C) **Analysis**

The contractor, owner, and separate contractor met early in the project to coordinate schedule requirements of allparties to satisfy the requirements of Section 6.1.1 of the A201. The separate contractor agreed to the contractor's approved project schedule for the phased turnover of rooms. A copy of this approved schedule is attached herein.

As noted in the contractor's attached as-built schedule, it turned out that the contractor started to turn over the rooms to the separate contractor as noted per the agreed-upon schedule, but then the contractor's finish work slowed for contractor-caused reasons, and as a result the separate FF&E contractor remained on site for an additional two months, which caused it to incur extended general conditions.

The separate contractor filed a claim with the owner for the amount of $XX for two months of extended general conditions. A copy of this claim is attached herein and the damages section of this claim evaluates the reasonableness of the charged amount. In addition, the delay section of this report provides a forensic schedule analysis that evaluates the two-month delay related to the contractor's finish work on the project. Lastly, attached is a copy of the check that the owner issued to the separate contractor as well as the separate contractor's release related to this payment. According to Section 6.2.3 of the A201, the owner requests reimbursement from the contractor because of this damage that the owner incurred due to the contractor's delays.

In the event that the contractor fails to promptly reimburse this amount to the owner, the owner will present this claim to the architect and will request that the architect withhold this amount from the contractor's future application for payment per Sections 9.4.1 and 9.5.1 of the A201. Specifically, Section 9.5.1.5 of the A201 allows for an offset due to damage caused by the contractor to the owner or a separate contractor.

D) **Conclusion**

The owner seeks reimbursement of $XX that represents the damage caused by the contractor's delinquent finish work to the owner's separate contractor, which the owner paid to the separate contractor.

## F.  Change Order Negotiation Issues

Standard contract forms provide the owner with a remedy in the event that the contractor and owner reach an impasse during change order negotiations. In such instances, the owner or its representative can typically direct the contractor to perform the work while the cost and time impacts are later worked out. In the event that an agreement is not reached, the contractor can trigger dispute resolution.

**Hypothetical Change Order Negotiation Issue: Change Order Negotiation Related to Roofing:** The owner retained the contractor under an AIA A101 agreement (design-bid-build with a stipulated sum) that incorporates the AIA A201 general conditions to construct a four-story multi-family apartment project. Shortly after the start of work the owner and architect decided to switch the exterior stucco cladding to a brick veneer. The architect believes this change should be a cost increase of $8 per square foot while the contractor feels it is an increase of $20 per square foot. As a result of the disagreement, the negotiation dragged on for a month or so and the contractor noted that it would not move forward with the change. Consequently, the owner and architect elected to issue the contractor a construction change directive (CCD) that directed the contractor to proceed with the work and noted the increase would be based upon the architect's estimated incremental difference in the square foot unit price between stucco cladding and standard brick veneer.

**Sample Entitlement Outline:**

A) **Conclusion**

The architect issues this CCD to the contractor that is signed by the owner that directs the contractor to proceed with the modification relating to the change from a stucco veneer to a standard brick veneer. The CCD notes no change to the contract time and a $XX lump sum increase to the contract sum. Please proceed with this work as required by contract.

B) **Rule**

Per Sections 7.3.1 and 7.3.2 of the A201, in the event of a lack of total agreement to the terms of a change order, the architect can prepare and the owner and the architect can sign a construction change directive that directs the contractor to proceed with a change in the work prior to agreement on adjustment, if any, in the contract sum or contract time, or both.

Section 7.3.3 notes that if the CCD provides for an adjustment to the contract sum, the adjustment shall be based on one of the following methods: (1) mutual acceptance of a lump sum; (2) unit prices stated in the contract documents or subsequently agreed upon; (3) cost to be determined in a manner agreed upon by the parties and a mutually acceptable fixed or percentage fee; or (4) as provided in Section 7.3.4.

Per Section 7.3.4 of the A201, if the contractor does not respond promptly or disagrees with the method for adjustment to the contract sum, the architect shall determine the adjustment on the basis of reasonable expenditures and savings of those performing the work attributable to the change, plus an amount for overhead and profit, as set forth in the agreement, or if no such amount is set forth in the agreement, a reasonable amount.

Section 7.3.6 of the A201 requires the contractor to promptly proceed with the changed work upon receipt of a CCD and advise the architect as to the contractor's agreement or disagreement with the method, if any, provided in the CCD for determining the proposed adjustment in the contract sum or contract time.

If the contractor disagrees with the architect's determination of the adjustment in the contract time or contract sum, it can trigger Article 15 dispute resolution.

C) **Analysis**

The architect hereby issues the contractor a CCD to proceed with the change from stucco veneer to brick veneer. A copy of the executed CCD form is attached. The CCD notes no adjustment in the contract time as the stucco veneer work is well off the critical path of the project as shown in the contractor's most recent project schedule, which is attached herein. The CCD adjustment in time is based on a lump sum of $XX, which is derived by multiplying the overall square footage of stucco by $8 per square foot, which is the premium that the architect calculates for this work. Backup to support these unit prices is also attached herein.

By contract (Section 7.3.6 of the A201), the contractor is required to proceed with this work and if it objects to the architect's position in terms of time or money, it may file a claim under Article 15 of the A201 (Sections 7.3.5 and 7.3.9). Also, the contractor may requisition for costs related to this CCD per Section 7.3.9 of the A201.

D) **Conclusion**

The attached CCD requires the contractor to proceed with the change from stucco to brick on the exterior veneer. This report attaches the architect's position and basis regarding the time and cost impacts of this change.

# IV. Typical Subcontractor Claims Against Contractors

The most common types of claims that subcontractors assert against contractors relate to design issues, administrative issues, owner or contractor performance issues, third party issues, or change order negotiation issues. In many instances, a subcontractor's claim relates to an owner issue, such as design, administration, or a performance issue—these claims are considered pass-through claims, as the contractor will pass the subcontractor's claim through to the owner to review.

Standard subcontract forms note that the subcontractor is bound to the owner's decision on the claim and the subcontractor shall pay for all costs the contractor incurs in the prosecution of the subcontractor's claim. This scenario is often contentious because a contractor inherently wants to maintain a positive relationship with the owner, and subcontractor claims often strain owner-contractor communications. Regardless, the contractor is obligated to pass through a subcontractor's claim if it believes the claim has merit and is not overstated, and if it refuses to do so for merely relationship reasons, the contractor may end up the liable party for the pass-through claim.

The following discussion assumes that the claimant is the subcontractor and the contractor-subcontractor agreement is an AIA A401 document that incorporates the A201 general conditions.

## A. Design Issues

Subcontractor claims related to design issues typically relate to the owner's design of the project. Thus, most subcontractor design claims are pass through in nature unless the contractor is responsible for design. As noted above, design claims generally involve the identification of a changed condition through a comparison of the work outlined in the contract documents and a changed condition, which affects the subcontractor's contract sum and/or contract time. The change might be an unanticipated subsurface condition, additional or changed work on an updated set of plans, a design omission that must be incorporated into the project, a specified product that is no longer available, or necessary work to address a design error. The subcontract agreement between the parties should detail the contract documents that the contractor is bound to, so this exercise is often straightforward.

**Hypothetical Added Scope Claim Scenario: Pass-Through Claim to the Owner:** The owner retained the contractor to construct a seven-story multi-family building with underground parking. The owner-contractor agreement is an AIA A133 agreement (CM at risk with a GMP) that incorporates the AIA A201 general conditions. After the owner and contractor agreed to a guaranteed maximum price (GMP), the designer materially revised the architectural plans in the basement area of the project. Specifically, the new plans note additional wall assemblies not found on the set of plans that the GMP is based upon. Shortly after the contractor transmitted the updated plans to the contractor's framing and drywall subcontractor, the subcontractor made a claim against the contractor for this additional work, which the contractor intends to pass through to the owner. An entitlement outline for this type of claim might be as follows.

**Sample Entitlement Outline:**

A) **Conclusion**

Designer's updated set of architectural plans include additional wall assemblies not shown on plans upon which the GMP is based. This represents additional work that will increase the framing and drywall subcontractor's contract sum by $XX; therefore, the subcontractor submits a claim for this additional work.

B) **Rule**

Per Section 1.3 of the A401, the A201 general conditions govern the subcontract agreement. Per Article 2 of the A401, for the AIA A201 general conditions, the contractor shall assume toward the subcontractor all obligations and responsibilities that the owner, under the A201, assumes toward the contractor, and the subcontractor shall assume toward the contractor all obligations and responsibilities that the contractor, under the A201, assumes toward the owner and the architect. Where A201 provisions are inconsistent with the subcontract, the subcontract agreement shall govern.

Section 5.2 of the A401 indicates that the contractor may order the subcontractor in writing to make changes in the work within the general scope of the subcontract, including additions, deletions, or other revisions. The subcontractor, prior to the commencement of such changed or revised work and if applicable, shall submit promptly to the contractor a written claim for an adjustment to the subcontract sum and subcontract time for such revised work in a manner consistent with the requirements of the subcontract documents.

Section 5.3 of the A401 notes the subcontractor shall make all claims promptly to the contractor for additional cost, extensions of time, and damages for delays, or other causes. A claim that becomes part of a claim made by the contractor to the owner shall be made not less than two working days preceding the time by which the contractor's claim must be made.

Per Sections 3.2.2–3.2.4 of the A201, before starting each portion of the work, the contractor shall review the contract documents and other information and shall issue an RFI to the architect regarding any nonconformity discovered by or made known to the contractor. If, upon review of the architect's response, the contractor feels it is entitled to additional cost or additional time, the contractor shall proceed to Article 15.

Sections 8.3.1–8.3.2 of the A201 note that if the contractor is delayed by: (1) an act or neglect of the owner or the architect; (2) changes ordered in the work; (3) delays beyond the contractor's and owner's control; (4) delay authorized by the owner pending mediation and binding dispute resolution; or (5) by other causes that the contractor asserts, the architect shall decide what time, if any, is owed to the contractor. If the contractor disagrees with the architect's position, it may make a claim per Article 15.

Per Section 15.1.3.1 of the A201, the contractor shall notice the owner, the IDM, and the architect within 21 days of the occurrence of the event giving rise to the claim or recognition of the event giving rise to the claim. Section 15.2.1 requires the contractor to refer the claim to the IDM for initial decision.

Article 6 of the A401 notes mediation is a condition precedent to binding dispute resolution, and the binding dispute resolution selection per Section 6.2 of the A401 is arbitration.

C) **Analysis**

The set of plans that the subcontractor based its bid upon, which are clearly indicated in the subcontract agreement, reflect 250 linear feet of interior wall assembly in the basement area. See below for a detailed takeoff of this calculation.

Last week, the contractor furnished the subcontractor the architect's updated set of plans for the project and directed the subcontractor to proceed with the work according to such plans. A copy of the contractor's transmittal is included herein. The architect's updated set of architectural plans include 550 linear feet of interior wall assembly throughout the basement area, as detailed below. Thus, the updated set of architectural plans include 300 linear feet of wall assembly that fall outside of the subcontractor's scope of work.

The subcontractor recognizes the contractor's right to order the subcontractor to make changes in the work under Section 5.2 of the A401. The subcontractor understands the contractor's transmittal to be such a direction regarding any additional work that is shown on the updated set of architectural plans. Per Section 5.3 of the A401, the subcontractor promptly makes this claim for the cost of performing an additional 300 feet of wall assembly work in the basement area of the project.

The subcontractor requests the contractor to issue notice to the owner regarding this claim under Section 15.1.3.1 and then to file this claim, with any applicable contractor markup, to the owner per Section 15.2.1 of the A201. In the event that the contractor elects not to pass this claim through to the owner, the subcontractor will trigger dispute resolution per Article 6 of the subcontract agreement.

D) **Conclusion**

In sum, the subcontractor asserts a claim to the contractor of $XX for the estimated cost to perform the added wall assembly work in the basement area of the project that is shown in the updated architectural drawings for the project that the contractor formally directed the subcontractor to follow.

## B. Administrative Issues

A subcontractor's claim regarding an administrative issue may relate to administrative tasks of the owner or the contractor. The impact of an owner's maladministration often affects subcontractor performance because RFIs, submittals, change orders, or inspection requests often involve the work of one or more subcontractors. A contractor may also be the cause of maladministration as it is responsible for passing through subcontractor RFIs, submittals, change order requests, etc. in a timely manner. Moreover, the contractor is responsible for making payments to subcontractors per the terms of the subcontract agreements. Administrative issues typically involve a pattern of delinquent behavior by the owner or contractor that gives rise to a subcontractor's claim.

**Hypothetical Administrative Issue: Contractor's Failure to Make Timely Payments to Subcontractor:** The owner retains the contractor under an AIA A102 agreement (design-bid-build with a GMP) for an industrial warehouse project. The contractor retained its subcontractors under A401 agreements. The contractor has consistently failed to pay the electrician per the terms of the A401 subcontract. On the most recent payment application, the contractor again did not pay the subcontractor within seven days upon receipt of associated funds from the owner, so the subcontractor issued the contractor a seven-day stop work notice. More than seven days expired from the date the contractor received this notice without payment being made so the electrical subcontractor stopped work on the project and demobilized. The contractor paid the subcontractor shortly after the electrician's demobilization, but the subcontractor seeks reimbursement for reasonable costs of demobilization, delay, and remobilization, which is the basis of the electrician's claim.

**Sample Entitlement Outline:**
  A) **Conclusion**
     The subcontractor makes this claim against the contractor for an increase in the subcontract sum of $XX and an increase in the subcontract time of XX, due to the contractor's continued maladministration of the payment provisions of the subcontract agreement.
  B) **Rule**
     Per Section 11.1.3 of the A401, the contractor shall pay the subcontractor no later than seven working days after the contractor

receives payment from the owner for funds related to the subcontractor's work. In the event the contractor does not pay the subcontractor through no fault of the subcontractor within this seven-day period, Section 4.8 of the A401 allows the subcontractor to stop work on the project until payment is received if the subcontractor notices the contractor of its right to stop work due to non-payment and seven days elapse from the contractor's receipt of this notification. Section 4.8 also notes that if the subcontractor properly halts work, the contractor shall modify the subcontract with the subcontractor's reasonable costs of demobilization, delay, and remobilization.

If the contractor fails to modify the subcontract upon request for reasonable demobilization, delay, and remobilization costs, Section 5.3 requires the subcontractor to promptly file a claim with the contractor that addresses these impacts. Because this type of contract does not pertain to an owner issue, there is no need to address any requirements under the A201 general conditions. In the event the contractor disregards or rejects this claim, the subcontractor may trigger dispute resolution via the requirements of Article 6 of the A401.

C) **Analysis**

As shown in the table below, the contractor has consistently breached Section 11.1.3 of the A401 that requires the contractor to pay the subcontractor within seven days of receipt of payment from the owner. After repeated written reminders to the contractor of this provision and its continual breaches (see attached emails), the subcontractor issued the contractor a formal seven-day notice per Section 4.8 of the A401 because seven days had elapsed since the contractor's receipt of the most recent payment application without payment to the subcontractor.

After seven additional days expired from the subcontractor's notice of lack of payment, the subcontractor halted work and demobilized from the project, as allowed by Section 4.8. See the attached letter from the subcontractor to the owner advising of this work cessation and demobilization. Shortly after the subcontractor's demobilization, the contractor made payment to the subcontractor for the outstanding monies. Per Section 4.8, the subcontractor hereby requests that the contractor issue a modification to the subcontract to cover the "reasonable costs of demobilization, delay, and remobilization." The subcontractor also requests a modification to the subcontract time to cover the period of time that elapsed from the subcontractor's stoppage of work to the subcontractor's remobilization.

Please refer to the delay and damages section of this claim that provides the basis for the subcontractor's time extension request and increase to the subcontract sum. The subcontractor promptly makes this claim per Section 5.3 of the A401. Because this issue does not involve the owner, there are no pass-through requirements that the subcontractor needs to meet.

D) **Conclusion**

The subcontractor requests that the contractor timely review and approve this claim for the cost and time impacts of the contractor's payment maladministration. Per Section 4.8, the subcontractor exercised its contract remedy to halt the work due to non-payment and the subcontractor is entitled to the time and cost impacts related to the subcontractor's demobilization, delay, and remobilization, which are outlined herein. In the event the contractor disregards or rejects this claim, the subcontractor intends to trigger dispute resolution via the requirements of Article 6 of the A401.

**Hypothetical Maladministration Claim: Late Submittal Review:**

The owner retained the contractor to construct a high-rise condominium project. The owner-contractor agreement is an AIA A102 agreement (design-bid-build with a GMP) that incorporates the AIA A201 general conditions. In turn, the contractor retained subcontractors under AIA A401 agreements, including an electrician for all electrical work required under the contract documents. During the construction phase of the project, the architect took an additional 60 days to review the electrician's light fixture submittal. This delay caused a one-month delay to the electrician's scope of work and the electrician will incur associated delay damages as a result. The electrician filed a notice of claim with the contractor in a timely manner, and now submits its formal claim to the contractor for submission to the IDM for an initial decision.

**Sample Entitlement Outline:**

A) **Conclusion**

The architect was two months late in its review of the electrician's light fixture submittal when compared to the required duration noted in the contractor's approved submittal schedule, which has caused a XX calendar day delay to the electrician's overall work schedule. The electrician seeks an equitable adjustment in the subcontract time for this amount and an equitable adjustment in the subcontract sum in the amount of $XX due to this issue.

B) **Rule**

Per Section 1.3 of the A401, the AIA A201 general conditions govern the subcontract agreement except to the extent of a conflict with the A401, in which case the A401 has priority. Per Article 2 of the A401, the contractor shall assume toward the subcontractor all obligations and responsibilities that the owner, under the A201, assumes toward the contractor, and the subcontractor shall assume toward the contractor all obligations and responsibilities that the contractor, under the A201, assumes toward the owner and the architect. Where A201 provisions are inconsistent with the subcontract, the subcontract agreement shall govern.

Section 3.10.2 requires the contractor to submit a submittal schedule for the architect's approval. If the contractor fails to submit a submittal schedule, or fails to provide submittals in accordance with the approved submittal schedule, the contractor shall not be entitled to any increase in contract sum or contract time based on the time the architect spends in reviewing submittals.

Per Section 4.2.7 of the A201, the architect shall return the contractor's submittals per the contractor's approved submittal schedule or, in the absence of an approved submittal schedule, with reasonable promptness.

If the contractor is delayed by an act or neglect of the owner or the architect, Section 8.3.1 of the A201 notes that the architect shall decide what time, if any, is owed to the contractor. If the contractor disagrees with the architect's position, it may make a claim per Article 15.

Per Section 5.3 of the A401, the subcontractor shall promptly file claims to the contractor. Per Sections 15.1.5 and 15.1.6 of the A201, the contractor shall issue notice of claims for cost and time, respectively, per Section 15.1.3 of the A201. Per Section 15.1.3, the contractor shall provide notice to the owner, the architect, and the IDM within 21 days of the event giving rise to the claim or recognition of the event giving rise to the claim. Once this notice is provided, the contractor shall then provide claim backup to the owner, the architect and the IDM and request an initial decision from the IDM. If the contractor fails or refuses to follow these procedures, the subcontractor can trigger mediation and binding dispute resolution per Article 6 of the A401.

C) **Analysis**

Here, the subcontractor previously provided timely notice of claim to the contractor regarding the architect's delays in returning submittals per Section 5.3 of the A401. In turn, the contractor provided proper and timely notice to the owner, the architect, and the IDM under Section 15.1.3.1 of the A201.

The subcontractor hereby issues this complete claim package to the contractor so that it can apply allowable markups to it and forward it to the owner, the architect, and the IDM to get an initial decision from the IDM. In the event the contractor does not request a decision from the IDM, the subcontractor shall trigger dispute resolution per Article 6 of the A401.

Per Section 4.2.7 of the A201, the architect shall review submittals per the contractor's approved submittal schedule. Per the approved submittal schedule, the architect had 15 days to review the contractor's lighting fixture submittal. Despite this requirement, the architect took 75 days to review and approve the submittal with no comments. Thus, the architect returned the submitted 60 days, or two months, late. As noted in the delay section of this report, this issue caused a one-month delay to the electrician's work. The electrician's delay damages related to this one-month impact are detailed in the damage section of this report.

D) **Conclusion**

In sum, the subcontractor files a claim to the contractor for impacts caused by the architect's late submittal review. The subcontractor requests a XX calendar day extension in the subcontract time and a $XX increase in the subcontract sum. As this claim is pass-through in nature, please forward this claim to the IDM for a decision pursuant to AIA A201 Section 15.2.1 and advise the subcontractor of the outcome upon receipt of the initial decision.

## C. Owner or Contractor Performance Issues

A large category of subcontractor claims relates to impacts caused by the performance of the owner or the contractor. When an owner's or contractor's performance interferes with the subcontractor's work, contractor claims often manifest. Because the contractor typically performs a far greater amount of work than the owner, subcontractor performance claims against the contractor are far more common. Some of the commonly noted interference claims made by subcontractors include stacking of trades, turnover of workforce, crew size inefficiency, dilution of supervision, errors and omissions, early occupancy, site access, logistics, fatigue, ripple, overtime, and season and weather changes. Proving entitlement for a performance claim is generally done through identifying when a contractor's performance is required against when it actually took place. Key documents often include the baseline schedule, daily reports, schedule updates, correspondence, and photographs.

**Hypothetical Performance Claim Issue: Delays by the Owner's Separate Contractor:** The owner retained the contractor to construct a hotel project. The owner-contractor agreement is an A101 agreement (design-bid-build delivery method) that incorporates the A201 general conditions. The owner also retains a separate contractor to complete demolition of an existing structure before the contractor can commence the foundation work. Within a week of the owner's notice to proceed, the contractor mobilized to the site as required per the prime contract. Upon receipt of the owner's notice to proceed, the contractor directed the foundation subcontractor to mobilize as well in order to meet the approved project schedule. The contractor's agreement with the foundation subcontractor was an AIA A401 form. The foundation contractor was unable to start its critical work upon mobilization because the owner's separate demolition contractor had failed to complete the demolition work, which caused the foundation contractor's equipment to sit idle for two weeks once on site, and it caused a critical path delay to the project of two weeks as well.

**Sample Entitlement Outline:**

  A) **Conclusion**

      The owner's separate contractor caused a two-week delay to the foundation subcontractor's work, as the owner's demolition work was a predecessor to the foundation subcontractor's work. The foundation subcontractor seeks an equitable adjustment in the subcontract sum of $XX and subcontract time of XX.

B) **Rule**

Per Section 5.3 of the A401, the subcontractor shall promptly file claims to the contractor. Per Sections 15.1.5 and 15.1.6 of the A201, the contractor shall issue notice of claims for cost and time, respectively, per Section 15.1.3 of the A201. Per Section 15.1.3, the contractor shall provide notice to the owner, the architect, and the IDM within 21 days of the event giving rise to the claim or recognition of the event giving rise to the claim. Once this notice is provided, the contractor shall then provide claim backup to the owner, the architect and the IDM and request an initial decision from the IDM. If the contractor fails or refuses to follow these procedures, the subcontractor can trigger mediation and binding dispute resolution per Article 6 of the A401.

The AIA A201 general conditions govern the subcontract agreement except to the extent of conflicts with the A401. Per Article 2 of the A401, for the AIA A201 general conditions, the contractor shall assume toward the subcontractor all obligations and responsibilities that the owner, under the A201, assumes toward the contractor, and the subcontractor shall assume toward the contractor all obligations and responsibilities that the contractor, under the A201, assumes toward the owner and the architect. Where A201 provisions are inconsistent with the subcontract, the subcontract agreement shall govern.

Section 6.2.3 of the A201 makes the owner responsible to the contractor for costs the contractor incurs due to a separate contractor's delays, improperly timed activities, damage to the work or defective construction. Per Section 8.3.1, if the contractor is delayed by an act or neglect of the owner or the architect, the architect shall decide what time, if any, is owed to the contractor. If the contractor disagrees with the architect's position, it may make a claim per Article 15.

C) **Analysis**

The subcontractor provided timely notice of claim to the contractor upon its mobilization to the site and recognition that it could not commence its critical work, and the contractor provided proper and timely notice to the owner, the architect, and the IDM

upon receipt of the subcontractor's notice. The foundation subcontractor issues this complete claim package to the contractor
so the contractor can include it in the contractor's claim to the
owner, the architect, and the IDM, so the contractor can get an
initial decision from the IDM. In the event the contractor does
not request a decision from the IDM, the subcontractor shall
trigger dispute resolution per Article 6 of the A401. Attached is a
copy of the notice issued by the subcontractor.

The project schedule attached to the prime contract notes that the
contractor and foundation subcontractor will mobilize to the site
within one week of the owner's notice to proceed to the contractor.
After the one-week mobilization period, the foundation subcontractor was to commence excavation work in the area of the previously demolished structure. Attached is a copy of the approved
project schedule and the contractor's directive to the subcontractor to mobilize to the site, which is attached to the owner's notice
to proceed to the contractor.

Attached are photographs of the site that confirm that the owner's
demolition contractor had yet to complete its work upon the
subcontractor's mobilization to the site, which caused a delay
to the subcontractor's work. Also included are photographs over
the two-week delay period. The reader is instructed to refer to
the delay section of the report for the schedule analysis and the
damages section of the report for the cost analysis.

D) **Conclusion**

The owner's separate contractor failed to complete demolition
work as required by contractor's schedule and this caused a delay
to the subcontractor's critical foundation work. Accordingly, the
subcontractor is entitled to an equitable adjustment in the subcontract time of XX calendar days and an equitable adjustment
in the subcontract sum of $XX.

**Hypothetical Performance Claim Issue: Delays by the Contractor:**
The owner retained the contractor to construct a 25-story hotel
project. The owner-contractor agreement is an A101 agreement
(design-bid-build with a stipulated sum) that incorporates the A201
general conditions. The contractor self-performed the concrete work

and it retained subcontractors for all other work under A401 agree-ments. Per the construction schedule, the metal framing was to start on level one once the structural post-tensioned concrete was installed up to level four. Thereafter, the concrete subcontractor was to cast a new level every two weeks. The project started out on schedule, but once the concrete contractor got up to Floor 4, it was unable to cast the higher floors every two weeks. On average, it took over three weeks to cast Floors 5 through 25, which caused a four-month delay to the framer's work.

**Sample Entitlement Outline:**

A) **Conclusion**

The subcontractor submits this claim to the contractor for an equi-table adjustment in the subcontract time of XX calendar days and an adjustment in the subcontract sum of $XX for associated delay damages relating to the contractor's delays related to the struc-tural concrete on the project.

B) **Rule**

Per Section 5.3 of the A401, the subcontractor shall promptly file claims to the contractor for additional cost, extensions of time, and damages for delays. Section 9.2.2 indicates that the subcontractor shall achieve substantial completion of the subcontractor's work pursuant to the contractor's approved schedule included in the prime contract agreement.

Dispute resolution provisions between the subcontractor and the contractor are defined in Article 6 of the A401, where mediation is noted as a condition precedent to the selected method of binding dispute resolution, which is arbitration in this instance.

C) **Analysis**

The subcontractor provided timely notice of the claim to the con-tractor shortly after the contractor failed to maintain its concrete schedule after Floor 4. A copy of this notice is attached herein.

The project schedule attached to the prime contract notes that the contractor will cast the concrete work on Floors 5 through 25 based on a two-week per floor duration. The concrete work is a predecessor activity to the metal framing work. From Floor 5 through 25, the contractor's average casting time for each floor was over three weeks, which caused a four-month delay to the framing subcontractor's work. Per Section 5.3 of the A401, the framing subcontractor issues this complete claim package to the contractor to detail the cost and time impacts associated with this four-month impact.

See the delay section for a complete schedule analysis of this four-month impact and see the damages section for the delay damages related to this impact. Also included are site photographs of the status of the concrete work on a weekly basis, which confirm the delays when compared to the approved project schedule that the subcontractor is bound to under Section 9.2.2 of the A401.

In the event that the contractor denies this claim, the subcontractor shall trigger dispute resolution per Article 6 of the A401, which mandates mediation as a condition precedent to arbitration.

D) **Conclusion**

The contractor's concrete delays caused a XX calendar day delay to the subcontractor work. Accordingly, the subcontractor is entitled to an equitable adjustment in the subcontract time of this amount and an adjustment in the contract sum of $XX for associated delay damages.

## D.  Force Majeure Issues

Force majeure issues result from impacts due to events that are beyond the control of the subcontractor, the contractor, and the owner, such as abnormal weather, governmental shutdowns, strikes, pandemics, etc. Standard subcontract forms may limit the remedy for claimants of force majeure impacts to time and not damages, because neither the claimant nor respondent has control over force majeure issues. Proving force majeure impacts generally involves establishing that: (1) a force majeure impact did in fact take place; (2) the impact affected the subcontractor's work; and (3) the subcontract allows the subcontractor to recover time and potentially damages under some subcontract agreements. Establishing force majeure impacts is often done through information from reliable outside sources that confirm the force majeure impact indeed took place and this is augmented by project information that confirm no work or limited work took place on the project as a result of this impact.

**Hypothetical Force Majeure Issue: Abnormal Weather:** The owner retains the contractor under an A102 agreement (design-bid-build with a GMP) that incorporates the A201 general conditions. The project involves the construction of high-rise mixed-use project that has a post-tensioned concrete frame. During the construction phase of work, the contractor's concrete work is slowed by abnormally wet weather during the months of April and May. The contractor retained the concrete subcontractor under an AIA A401 agreement. The concrete subcontractor seeks an extension of time due to the abnormal weather. The concrete work falls on the critical path of the project, so the overall substantial completion date of the project slipped as well. Thus, the contractor is going to pass this claim through to the owner.

## Sample Entitlement Outline:

A) **Conclusion**

Abnormal weather impacted the subcontractor's concrete work in April and May. Accordingly, this represents the subcontractor's formal claim to recover the associated time impact of XX calendar days.

B) **Rule**

Per Section 5.3 of the A401 subcontract, the subcontractor shall promptly file claims to the contractor for additional cost, extensions of time and damages for delays, or other causes. For subcontractor claims that become part of a claim made by the contractor to the owner, the subcontractor shall issue the claim not less than two working days preceding the time by which the contractor's claim must be made. Section 1.3 of the A401 incorporates the A201 general conditions except to the extent that the A201 conflicts with the terms of the A401. Where conflicts exist, the A401 governs.

Per Article 2 of the A401, the subcontractor shall apply the A201 general conditions such that the contractor shall assume toward the subcontractor all obligations and responsibilities of the owner and the subcontractor shall assume toward the contractor all obligations and responsibilities that the contractor assumes toward the owner and the architect. Where A201 provisions are inconsistent with the subcontract, the subcontract agreement shall govern.

Section 8.3.1 of the A201 general conditions notes that if the contractor is delayed at any time in the commencement or progress of the work by adverse weather conditions and the architect agrees with such delay, then the contract time shall be extended for such reasonable time as the architect may determine. Section 8.3.2 notes that all claims relating to time shall be made in accordance with applicable provisions of Article 15, and Section 8.3.3 confirms that Section 8.3 does not preclude recovery of damages for delay.

Article 15 of the A201 covers "claims and disputes." The first section of Article 15, Section 15.1.1, defines a claim as a demand or assertion by one of the parties seeking money and/or time per the terms of the contract. Section 15.1.6.2 specifically addresses "adverse weather" claims and notes that such claims shall be documented by data substantiating that weather conditions were abnormal for a period of time, could not have been reasonably anticipated, and had an adverse effect on the scheduled construction.

Moreover, per Sections 15.1.5 and 15.1.6, contractor claims for cost and time shall be noticed to the owner, the architect, and the IDM per Section 15.1.3, which requires notice within 21 days of the occurrence of the event giving rise to the claim or recognition of the event giving rise to the claim. Thus, subcontractors must provide the contractor with notice within 19 days of the occurrence of the event giving rise to the claim or recognition of the event giving rise to the claim.

C) **Analysis**

The subcontractor issued the contractor formal written notice within one week of recognition of the abnormal weather, immediately after the adverse weather of April and May had ceased. Attached is a copy of this letter. Thus, the subcontractor properly noticed this claim pursuant to Section 5.3 of the A401 and Section 15.1.3 of the A201.

The subcontractor requests that the contractor pass this claim through to the owner, the architect, and the IDM and request an initial decision from the IDM. As required by Section 15.1.6.2, this claim for impacts due to adverse weather conditions is supported by data substantiating that the weather conditions were abnormal for the noted time frame, such weather could not be reasonably anticipated, and the adverse weather had an adverse effect on the subcontractor's concrete work.

The project is located in Dallas, Texas. According to historical NOAA data, the average precipitation for the months of April and May for the past 60 years is 3.36 and 4.83 inches, respectively. The actual precipitation for April and May of this year was 5.56 and 16.96 inches, respectively. Hence, April was 1.65 wetter times the norm and May was 3.51 wetter times the norm. In sum, the weather in April and May was abnormal.

Also, the average days of rain in Dallas for the month of April is 7 and for the month of May is 10. In order to account for weekends, the average weather days per weekday equates to: April = [5 workdays / 7 day week] x 7 days of rain in April = 5; May = [5 workdays / 7 day week] x 10 days of rain in April = 7.14 (round down to 7). This past April it rained 10 workdays and this past May it rained 14 workdays. Thus, the weather was abnormal for 5 workdays in April and 7 workdays in May, for a total workday impact of 12 days. When you convert the 12 workdays to calendar days, it equates to 17 calendar days (12 workdays / 5 workdays per week * 7 calendar days per week = 16.8 calendar days; round up to 17 calendar days).

It is reasonable for concrete subcontractors to anticipate normal weather conditions. It is not common or reasonable for subcontractors to anticipate weather conditions that are above or below the norm.

Attached is a table that identifies the 24 workdays where it rained at the site in April and May. The subcontractor's crews did not work on the site during these days, but the subcontractor's supervisory team was on site during this time. Attached are the subcontractor's daily reports that were previously submitted to the contractor that confirm no physical work took place during these 24 rain days. Also included are photographs of these dates from the project's site camera that confirm no concrete work took place. As noted above, when you deduct the 12 average rain days during the April to May timeframe from the actual figure of 24, it equates to an impact of 12 workdays, or 17 calendar days.

D) **Conclusion**

The subcontractor is entitled to a XX calendar day extension based on the abnormal rain delays that took place at the site in April of May. These days were abnormal, unforeseeable, and prevented concrete work at the site.

## E. Change Order Negotiation Issues

Nearly all construction projects involve additive or deductive change orders between the owner and the contractor, and the contractor and the subcontractors. When the parties agree that a change order is owing, proving entitlement is straightforward, as a change order represents recognition of a cost and/or time impact to the contractor's work, which often involves the subcontractors' work. Change order negotiation disputes between the contractor and subcontractors often relate to the magnitude of the impact on the subcontractor's work, which brings the negotiation to a stalemate. Thus, the bulk of effort for this type of claim relates to the delay and damages section of the claim; however, an entitlement review is still necessary.

**Hypothetical Change Order Negotiation Issue: Failed Change Order Negotiation:** The owner retains the contractor under an AIA A102 agreement (design-bid-build with a GMP) for an industrial warehouse project. The owner, through its architect, sought to add mechanical work to the contractor's scope of work through the traditional negotiation process. The contractor solicited pricing related to this change from its subcontractors. The owner did not agree with the cost and time impacts provided by the mechanical subcontractor, which represented the bulk of the change order. Consequently, the architect issued the contractor a construction change directive (CCD) for the mechanical work, so the contractor directed the mechanical subcontractor to complete the changed work. Thereafter, the contractor provided the owner and the architect with backup related to the changed work, but the parties still could not reach an agreement on the cost and time impacts so the architect issued an interim determination for the change. Because the contractor and mechanical subcontractor disagreed with the architect's interim agreement position, it triggered dispute resolution under Article 15 of the A201.

**Sample Entitlement Outline:**

A) **Conclusion**

The mechanical subcontractor disagrees with the architect's interim determination relating to the CCD for additional mechanical work on the project. Accordingly, the subcontractor requests that the contractor request an interim decision from the

IDM under Section 15.2.1 of the A201 general conditions. The subcontractor seeks an equitable adjustment in the subcontract time of XX calendar days and an increase in the subcontract sum of $XX.

B) **Rule**

Per Section 5.1 of the A401 agreement, the owner may make changes to the contractor's work by issuing modifications to the prime contract. Section 1.3 of the A401 incorporates the A201 general conditions except to the extent that the A201 conflicts with the terms of the A401. Where conflicts exist, the A401 governs.

Section 7.1 of the A201 allows the owner to make modifications to the prime contract via a change order, CCD, or an order for a minor change in the work. Section 5.2 of the A401 allows the contractor to order the subcontractor to make changes in the work per owner-ordered modifications directed under the prime contract. When this happens, Section 5.2 requires the subcontractor to promptly submit a claim for changes to the subcontract sum and subcontract time caused by the modification. Moreover, Section 5.3 of the A401 notes that if a subcontractor claim becomes part of a claim made by the contractor to the owner, the subcontractor shall issue the claim not less than two working days preceding the time by which the contractor's claim must be made.

Per Sections 7.3.5 and 7.3.9 of the A201, if the contractor disagrees with the architect's adjustment in the contract time and contract price related to a CCD, the contractor may pursue a claim per Article 15. Article 15 of the A201 covers "claims and disputes." Sections 15.1.5 and 15.1.6 of the A201 note that contractor claims for cost and time shall be noticed to the owner, architect, and IDM per Section 15.1.3, which requires notice within 21 days of the occurrence of the event giving rise to the claim or recognition of the event giving rise to the claim. Thus, to comply with Section 5.3 of the A401, the subcontractor shall provide the contractor with notice within 19 days of the occurrence of the event giving rise to the claim or recognition of the event giving rise to the claim. After the claim is properly noticed, the contractor can submit its formal claim and request an initial decision from the IDM under Section 15.2.1 of the A201.

C) **Analysis**

The subcontractor disagrees with the architect's position on both time and cost associated with the mechanical-related CCD. The subcontractor issued the contractor formal written notice of claim within ten days of the architect's CCD to the contractor, and within five days of the contractor's direction to the subcontractor to proceed with the modification work. Accordingly, the subcontractor has met its notice requirements under the A401 and A201 forms.

The architect has directed the contractor to modify the contract through a CCD, and the contractor has in turn directed the subcontractor to proceed with the CCD. Thus, the architect acknowledges that a change to the contract sum and/or contract time is owing, but the amount of such changes has not yet been agreed upon, hence the issuance of a CCD rather than a traditional change order. The subject of this CCD relates to mechanical changes so the parties agree that the mechanical subcontractor is entitled to a change in the subcontract. The issue at bar is the amount of change to the subcontract sum and subcontract time. Please refer to the delay section of this report for the subcontractor's window's analysis that establishes that the subcontractor is entitled to the noted time extension and refer to the damages section of this report that establishes the amount that the subcontractor is due for delay damages and direct costs related to this modification.

D) **Conclusion**

The subcontractor disagrees with the architect's position on the impacts to the contract sum and contract time due to the CCD related to mechanical modifications. The subcontractor therefore requests the contractor request an initial decision from the IDM on this matter. The subcontractor seeks an equitable adjustment in the subcontractor time of XX calendar days and an increase in the subcontract sum of $XX.

Should the IDM reject this claim, or should the contractor fail to timely request this initial decision, the subcontractor will commence dispute resolution under Article 6 of the A401 by initiating mediation.

# V. Typical Contractor Claims Against Subcontractors

The most common types of claims that contractors assert against subcontractors relate to quality issues, schedule issues, administrative issues, design issues, third-party issues, and change order negotiation issues. The following discussion assumes that the claimant is the contractor and the contractor-subcontractor agreement is an AIA A401 agreement that incorporates the A201 general conditions.

## A. Quality Issues

Standard subcontract forms require the subcontractor to install work in accordance with the contract documents. When subcontractors fail to meet this requirement, contractor remedies include the stoppage of the work, withholding earned funds, or correction of the work after proper notice. In order to prove entitlement for a quality claim, the contractor must compare the installed work with the requirements of the contract documents or approved submittals and clearly note a difference between the two in order to evidence that the subcontractor's work falls below the contract requirements. In addition, contractors often cite applicable code provisions, manufacturer's recommendations, or industry standard publications for baseline quality requirements for specific divisions of work if the contract documents are silent on the portion of work in question and the subcontractor failed to issue an RFI to seek clarification.

**Hypothetical Subcontractor Quality Issue: Lack of Open Head Joint Weeps in the Brick Veneer:** The owner retained the contractor under an AIA A101 agreement (design-bid-build with a stipulated sum) to construct a four-story office building that is clad with brick veneer. The contractor retained its subcontractors under A401 agreements. After approximately a third of the brick veneer had been installed by the masonry subcontractor, the contractor noticed that the mason had failed to install open head joint weeps above all through wall flashing in the brick veneer, which is clearly noted in the architectural plans and described in the masonry specifications. Per the architectural plans, open head joint weeps are required every 24 inches on center. The masonry work and related flashing work are included in the masonry subcontractor's scope of work. Accordingly, the contractor rejects the work and notices the subcontractor of the remedies that the contractor maintains under the subcontract agreement for defect work issues.

**Sample Entitlement Outline:**

A) **Conclusion**

The masonry subcontractor failed to install open head joint weeps every 24 inches on center above the through wall flashing throughout the exterior brick veneer installed to date, as required by the architectural drawings and project specifications to which the subcontractor is bound to follow. Accordingly, the contractor will withhold funds from the subcontractor's payment application in the estimated amount to cure this defect and this letter serves as notice to the subcontractor that the contractor might arrange for this work to be corrected by others if the subcontractor fails to commence and continue corrective efforts within five working days upon receipt of this notice.

B) **Rule**

Per Section 4.6.1 of the A401, the subcontractor warrants that the subcontractor's work will conform to the requirements of the subcontract documents and will be free from defects. Article 8 of the A401 notes that the subcontractor's work includes all masonry work on the project per the contract documents, including all related weeps and flashings that are incorporated into the masonry work, or directly above or below the masonry work.

Section 1.1 of the A401 confirms the "subcontract documents" include the prime contract, which consists of the agreement between the owner and contractor and the other contract documents enumerated therein. In the event that the subcontractor's work does not conform to the prime contract, Section 4.2.5 of the A401 provides the contractor and the architect with the authority to reject the work of the subcontractor.

The contractor has several remedies if the subcontractor installs defective work. First, the contractor can reduce the subcontractor's earned progress payment for defective work under Sections 11.1.7.2.2 and 11.1.7.2.3 of the A401, which refer to Article 9 of the A201. Section 9.5.1.1 of the A201 allows the owner to withhold funds against the contractor for defective work not remedied.

Section 1.3 of the A401 incorporates the A201 into the subcontract documents and Article 2 of the A401 allows the contractor to withhold funds against the subcontractor for defective work, just as the owner can withhold funds against the contractor for defective work. The withholding for defective work should represent the estimated cost to remedy the defect.

Second, Section 4.6 of the A401 allows the contractor to arrange for the correction of the subcontractor's defective work and backcharge all costs against the subcontract if the subcontractor fails within five working days after receipt of notice from the contractor to commence and continue correction of the defective work with diligence and promptness. If the cost of correction exceeds the subcontractor's unpaid contract balance, the subcontractor shall pay the difference to the contractor.

C) **Analysis**

The architectural drawings and project specifications denote open head joint weeps every 24 inches on center above all through wall flashing integrated within or below all masonry work on the project. Screenshots of these requirements can be found below.

The masonry subcontractor has failed to install any weeps above the through wall flashing installed to date, which renders the flashing useless. Below are numerous photographs that confirm this defect exists. The subcontractor is directed to install the weeps as required on the remaining brick masonry veneer work, and is directed to remedy this defect on all the installed brick masonry veneer work.

Per Sections 11.1.7.2.2 and 11.1.7.2.3 of the A401, the contractor intends to offset the subcontractor's payment application with the estimated cost to correct this defect. See the damages section of this report for a breakdown of this amount. Moreover, per Section 3.5 of the A401, this letter provides formal notice of the masonry subcontractor of the contractor's right to take over this corrective effort if the subcontractor fails to initiate and continue correction of this work with diligence and promptness. Any costs incurred by the contractor for this effort will be backcharged to the subcontractor under Section 3.5 and if the cost of this effort exceeds the subcontractor's unpaid contract balance, the subcontractor shall pay the contractor this amount.

D) **Conclusion**

The brick masonry work is defective because the masonry subcontractor failed to install open head joint weeps every 24 inches on center above all through wall flashing integrated within or below all masonry work installed on the project to date. The contractor will offset the subcontractor's payment applications for the estimated amount of this corrective work as defined in the damages section of this report. Moreover, this letter serves as the contractor's formal notice to the subcontractor that the contractor may arrange for the correction of this work by retaining a supplemental masonry crew if the subcontractor fails to commence and continue correction of this work within five working days of the subcontractor's receipt of this letter.

## B. Schedule Issues

Subcontract agreements typically require the subcontractor to achieve substantial completion of the subcontract work pursuant to a defined date or per the approved project schedule that the contractor is bound to under the prime agreement with the owner. When a project is delayed for subcontractor-caused reasons, contractors often assert delay claims against the subcontractor for owner-pass-through claims and actual delay damages incurred by the contractor. In order to prove a subcontractor-caused delay, the contractor can compare the requirements of the baseline schedule (with excusable time extensions) or recent schedule updates with the as-built construction progress to identify subcontractor-caused delays in the work.

**Hypothetical Subcontractor Schedule Issue: Critical Path Delay:**

The owner retained the contractor under an AIA A101 agreement (design-bid-build) that incorporates the AIA A201 general conditions to construct a 30-story office building that has a steel frame superstructure. The A101 agreement stipulates a liquidated damage rate of $10,000 per calendar day. The contractor retains all of its subcontractors under AIA A401 agreements. The contractor's schedule started to slip because the steel subcontractor was unable to mobilize sufficient workers to the project for no purported reason. Twelve months after the owner issued the contractor a notice to proceed, the substantial completion date of the overall 36-month project schedule had slipped three months.

The owner asserted liquidated damages against the contractor, so the contractor intends to pass the liquidated damage assessment along to the steel subcontractor.

**Sample Entitlement Outline:**

A) **Conclusion**

The structural steel subcontractor is not maintaining the project schedule as required by the A401 agreement. The steel work falls on the critical path of the project and as a result of the steel subcontractor's slow progression of work, the substantial completion has, to date, slipped by three months and as a result the architect has withheld three months of liquidated damages against the contractor. Consequently, the contractor intends to offset the subcontractor's payment application by this assessed liquidated damage amount, as allowed per the terms of the A401 agreement.

B) **Rule**

Section 9.2.2 of the A401 requires the subcontractor to complete the structural steel work per the approved contract schedule that is an attachment to the prime contract and is a subcontract document per Section 1.1 of the A401 agreement. Sections 1.1 and 1.3, as well as Article 2 of the A401 also incorporate the A201 as a subcontract document.

If the subcontractor fails to achieve substantial completion of its work per Section 9.2.4, the contractor shall assess liquidated damages as set forth in Section 3.4. Section 3.4.1 notes that if the prime contract provides for liquidated damages, an assessment against the subcontractor shall only be to the extent caused by the subcontractor. Section 4.5 of the A101 agreement between the owner and the contractor stipulates the liquidated damage rate. In addition to liquidated damages, Section 3.4.2 of the A401 also permits the contractor to assert "costs of services or materials provided" due to the subcontractor's failure to execute the work so long as the contractor provides a seven-day notice to the contractor regarding what is to be provided and written compilations to the subcontractor of the services and materials provided by the contractor no later than the 15th day of the month following the contractor's providing such services or materials.

Section 11.1.7.2 of the A401 permits the contractor to reduce the subcontractor's earned payment application with amounts that are allowed under Article 9 of the A201. Section 9.5.1 of the A201 allows withholdings for, among other things, actual or liquidated damages associated with the anticipated delay and costs associated with repeated failure to carry out the work in accordance with the contract documents.

Lastly, Section 15.1.1 of the A201 general conditions does not require the owner to file a formal claim in order to impose liquidated damages. Thus, per Article 2 of the A401, the same holds true for the contractor's assessment of liquidated damages against the subcontractor.

C) **Analysis**

The subcontractor's structural steel work falls on the critical path of the project schedule. Section 9.2.2 of the A401 requires the subcontractor to complete its work per the approved project schedule. To date, the subcontractor's work is three months late, and this has caused a three-month delay to the substantial completion date of the project. The subcontractor caused these delays due to poor productivity and a lack of workers to meet the project schedule. Attached is the baseline schedule that is a subcontract document, as well as all schedule updates that have been previously provided to the subcontractor that note the continued slippage of the critical structural steel work. Also see the delay analysis of this report for a detailed schedule analysis.

As shown in the contractor's most recent payment application signed off by the architect, the architect has withheld liquidated damages against the contractor in the amount of $900,000 for the 90-day delay to the substantial completion date of the project. Per Section 3.4.1 of the A401, the contractor hereby passes through this full liquidated damage assessment amount to the subcontractor because the subcontractor is solely responsible for this delay. Note the liquidated damage amount of $10,000 per calendar day is stipulated in Section 4.5 of the A101 agreement between the owner and contractor. This prime contract is a subcontractor document per Article 1 of the A401. Although Section 3.4.1 of the A401 is silent on this issue, the contractor shall only assess liquidated damages to the extent that the owner has assessed or will likely assess them against the contractor. The liquidated damage calculation to date is detailed in the damages section of this report.

Per Section 11.1.7.2 of the A401, the contractor will reduce the subcontractor's earned payment application by this liquidated damage assessment amount. Note that Section 11.1.7.2 refers to Article 9 of the A201, which explicitly allows for this type of withholding. Moreover, per Section 3.4.2 of the A401, the contractor hereby provides the subcontractor with a seven-day notice that it plans on mobilizing two additional supervisors to the project to focus exclusively on this remaining structural steel work and intends asserting these charges against the subcontractor. The contractor will provide the subcontractor with a written compilation of these charges related to these additional supervisory efforts by the 15th of the month following the month that the contractor provides such services.

D) **Conclusion**

To date, the subcontractor has delayed the substantial completion date of the project by three months. The subcontractor is solely responsible for this delay. The owner has reduced the contractor's most recent payment application by $900,000 for the accrual of liquidated damages for this delay period. The contractor hereby passes this liquidated damage assessment through the subcontractor per the terms of the A401 agreement. This also serves as the contractor's seven-day notice per Section 3.4.2 of the A401 that it intends to mobilize two additional supervisors to the project to exclusively focus on this structural steel work.

## C. Subcontractor Administrative Issues

Contractors often withhold or backcharge earned funds from subcontractors due to subcontractor administrative issues such as the failure to clean up or the failure to make timely payment to sub-subcontractors or vendors for work performed and previously paid for. Proof of entitlement for this type of claim can be done by detailing the subcontract requirement that outlines the subcontractor's administrative duty and then providing evidence that confirms the subcontractor breached this duty, such as photographs of an unkempt jobsite or payment claims by sub-subcontractors or suppliers.

**Hypothetical Subcontractor Administrative Issue: Failure to
Clean Up:** The owner retained the contractor under an AIA A133
agreement (CM at risk with a GMP) that incorporates the AIA A201
general conditions to construct a data center. The contractor retained
all subcontractors via an AIA A401 form of agreement. During the
construction phase of the project, the contractor noticed the framing
and drywall subcontractor of its failure to clean up. The contractor later
arranged for another party to clean up the subcontractor's work and
asserted a backcharge against the subcontractor for the costs associated
with this effort.

**Sample Entitlement Outline:**

A) **Conclusion**

The subcontractor is obligated to keep the premises and surrounding area free from accumulation of waste materials and
rubbish. The subcontractor failed to meet this requirement. The
contractor noticed the subcontractor of this breach and took
photographs before and after the notice period to confirm that
the subcontractor took no action on this issue. As a result, the
contractor arranged for this cleanup to take place and now seeks
reimbursement per the A401 agreement.

B) **Rule**

The A401 contemplates the scenario where the subcontractor fails
to properly clean up during the execution of the subcontractor's
work. Section 4.5.1 notes that the subcontractor shall keep the
premises and surrounding area free from accumulation of waste
materials or rubbish caused by operations performed under the
subcontract. Section 4.5.2 indicates that if the subcontractor fails
to clean up provided for in the subcontract documents (namely
Section 4.5.1), the contractor may charge the subcontractor for the
subcontractor's appropriate share of cleanup costs in accordance
with Section 3.4.2.

Section 3.4.2 of the A401 outlines the backcharge procedure in the
event that the contractor has a claim against the subcontractor for
services or materials provided due to the subcontractor's failure
to execute the work. First, the contractor must provide the subcontractor with a seven-day notice prior to providing the service
or material, except in an emergency, in which case no notice is
required. Second, the contractor must provide the subcontractor

with a written compilation of the services and materials provided by the 15th day of the month that the contractor provides such services or materials. The subcontractor shall then reimburse the contractor for these costs.

If the subcontractor fails to reimburse the contractor for the cleanup costs, Section 3.5 provides the contractor with the remedy to offset the subcontractor's payment application with claims relating to the subcontractor's failure to carry out the work per the agreement, just as the architect has the right to offset the contractor's payment application per Article 9 of the A201 general conditions.

C) **Analysis**

The subcontractor has failed to keep the premises and surrounding area free from accumulation of waste materials or rubbish caused by the subcontractor's work, as required by Section 4.5.1 of the A401 and as evidenced by the photographs attached herein. The photographs are dated and list the location of the unkempt area. Per Section 3.4.2 of the A401, the contractor properly noticed the subcontractor of this breach in a letter dated May 15th, which also included these photographs. After seven days lapsed with no action by the subcontractor (confirmed by that attached photographs that were taken after day seven), the contractor engaged a cleanup company to cure this issue, which is an allowable contractor remedy under Section 3.4.2. As required by Section 3.4.2, please find attached a timely and written submission of the costs associated with this cleanup effort that took place last month. The contractor requests immediate reimbursement of this amount and if no such reimbursement is made, the contractor will withhold this amount from the subcontractor's next progress payment pursuant to Section 3.5 of the A401.

D) **Conclusion**

The subcontractor has failed to clean up its work per the terms of the A401 agreement. Accordingly, the contractor properly noticed the subcontractor of this issue and after the subcontractor failed to remedy this issue after the seven-day cure period, the contractor engaged a company to properly clean up the subcontractor's work. The contractor seeks reimbursement for these cleanup charges. If the subcontractor fails to make such reimbursement, the contractor will offset this amount from the subcontractor's next payment application, which is allowed per the A401 agreement.

## D. Subcontractor Design Issues

Contractors typically retain subcontractors to address delegated design portions of work such as fire suppression systems, shoring, metal framing, precast, etc. If a subcontractor's design does not adhere to the design criteria or performance requirements outlined in the contract documents and the owner asserts a design error or design omission claim against the contractor, the contractor in turn passes these types of claims through to the responsible subcontractor. For a contractor to prove entitlement for such a pass-through claim, the contractor should compare the design criteria per contract against the subcontractor's design to identify design elements that fall short of the mandated prescriptive criteria. Or, if the contract documents outline a performance criteria, the contractor shall contrast the performance criteria with the actual performance of the system designed and installed by the subcontractor to identify performance that falls below the requisite requirements.

**Hypothetical Subcontractor Design Issue: Design Error on Fire Suppression Plans:** The owner retained the contractor under an AIA A133 agreement (CM at risk with a GMP) that incorporates the AIA A201 general conditions to construct an industrial building. The construction documents delegate the design of the fire suppression system to the contractor. The contractor, in turn, retained a fire suppression subcontractor to perform this work. The contractor retained all of its subcontractors via A401 agreements. Specifically, the contract drawings note, "Provide a complete wet-pipe automatic sprinkler system throughout the project area. The system shall be designed per NFPA-13 light hazard." As the project neared substantial completion, the architect determined that the fire suppression system, as installed, failed to meet several requirements of NFPA-13 in terms of main size, water treatment, and foundation penetrations. The contractor passed this claim down to its fire suppression subcontractor.

**Sample Entitlement Outline:**

A) **Conclusion**

The architect notified the contractor that certain aspects of the fire suppression subcontractor's design and installation failed to meet NFPA-13 light hazard requirements in terms of the minimum diameter of the private services main, the water treatment system, and the size of the foundation penetrations. The architect indicates that these design and installation flaws will detrimentally affect the performance of the fire suppression system. Contractor hereby passes this claim through to the fire suppression subcontractor that designed and installed this system.

B) **Rule**

Article 8 of the A401 requires the subcontractor to execute all its work per the subcontract documents, which includes all labor, material, equipment, services, and other items required to complete such work. Section 1.1 of the A401 notes the subcontract documents include the contract documents for the project.

Section 4.9.2 of the A401 indicates that the subcontractor shall provide all professional design services required by the subcontract documents. It further notes that the contractor shall provide the subcontractor with all performance and design criteria outlined in the prime contract. Section 4.9.4 confirms that the subcontractor is entitled to rely upon the performance and design criteria received from the contractor. Section 4.9.5 allows the contractor to rely upon the adequacy, accuracy, and completeness of the design services provided by the subcontractor.

Section 3.3.5 requires the contractor to promptly notify the subcontractor of any fault or defect in the subcontractor's work or nonconformity with the subcontract documents. Under Section 4.6.1, the subcontractor warrants that its work will conform to the requirements of the subcontract documents and will be free from defects and that work not conforming to these requirements may be considered defective.

In the event that subcontractor's work is defective, Section 11.1.7.2 allows the contractor to reduce the subcontractor's payment application for defective work as provided for in Article

9 of the A201. Section 9.5.1 of the A201 allows the architect to withhold its certification for payment in whole or in part because of defective work not remedied.

C) **Analysis**

The contract documents require the design of the fire suppression system to conform to NFPA-13. Below is a screenshot of this requirement. NFPA-13 lists the minimum requirements for main size, water treatment, and foundation penetrations. A copy of NFPA-13 is attached herein.

The subcontractor's work, as defined in Article 8 of the A401, includes the design and construction of the fire suppression system per the contract documents. Section 1.1 of the A401 incorporates the contract documents into the subcontract documents. The architect has placed the contractor on notice that the subcontractor's fire suppression system fails to adhere to NFPA-13 requirements. Attached is the architect's position as well as photographs that evidence non-conformance in the subcontractor's work. The architect intends to withhold funds from the contractor related to this defect until it is cured.

Per Section 3.3.5 of the A401, the contractor hereby notifies the subcontractor of the noted defects in its fire suppression system. The subcontractor's fire suppression design does not conform to the requirements of the subcontract documents, which renders it defective per Section 4.6.1 of the A401. The contractor will exercise its right under Section 11.1.7.2 and will reduce the subcontractor's payment application as a result of this defective work, as provided for in Article 9 of the A201.

Please promptly commence the remedy of this defect.

D) **Conclusion**

The subcontractor's design of the fire suppression system is defective for failure to adhere to NFPA-13, which is an explicit design criterion. The contractor plans on withholding funds associated with this defect until it has been cured by the subcontractor. Please commence the remediation of this defect promptly.

## E. Subcontractor-Caused Interference

Standard subcontract agreements require the subcontractor to cooperate with the contractor in scheduling and performing the subcontractor's work to avoid conflict, delay in, or interference with the work of other subcontractors or separate contractors. When a subcontractor fails to adhere to this requirement and it results in additional costs to the contractor, the contractor often backcharges this cost to the responsible subcontractor. To establish entitlement for this type of claim, a contractor must prove that the responsible subcontractor failed to adhere to the subcontractor requirements and this failure caused damage to another subcontractor or a separate contractor, for which the contractor is now responsible.

**Hypothetical Subcontractor Interference Issue: Subcontractor Delays Cause Impacts to Separate Contractor:** The owner retained the contractor under an AIA A133 agreement (CM at risk with a GMP) that incorporates the AIA A201 general conditions to construct a high-rise hotel project. Per the terms of the agreement the owner will retain a separate contractor for all FF&E work. The contractor and the separate contractor agreed to coordinate the work per the contractor's schedule. Similarly, the contractor's subcontractors are required to coordinate with separate contractors. The contractor retained its subcontractors under AIA A401 forms. As it turns out, the contractor delivered many of the hotel rooms to the separate contractor several months late due to the slow work of its painting subcontractor, which increased the separate contractor's time on the project. As a result, the separate contractor issued the owner a claim for two months of extended general conditions, which the owner paid and then backcharged to the contractor. The contractor, in turn, backcharged this amount to its painting subcontractor. Note that the painting work was not on the critical path of the overall project, which ran through the contractor's exterior civil work. Thus, the painter's delay did not extend the substantial completion date of the overall project.

**Sample Entitlement Outline:**

A) **Conclusion**

Because of the slow progress of the painting subcontractor, the owner paid its separate FF&E contractor two months of extended general conditions. The owner then backcharged this amount to the contractor. The contractor hereby asserts this same backcharge against the painting subcontractor per the terms of the A401 agreement.

B) **Rule**

Section 4.2.2 of the A401 requires the subcontractor to cooperate with the contractor in scheduling and performing the subcontractor's work to avoid conflict, delay in, or interference with the work of the contractor, other subcontractors, the owner, or separate contractors.

Section 9.2 of the A401 notes that the subcontractor shall complete the work in accordance with the approved project schedule. Section 3.3.5 requires the contractor to promptly notify the subcontractor of any fault or defect in the subcontractor's work or nonconformity with the subcontract documents.

In the event that the subcontractor's work is defective, Section 11.1.7.2 allows the contractor to reduce the subcontractor's payment application for defective work as provided for in Article 9 of the A201. Section 9.5.1 of the A201 allows the architect to withhold its certification for payment in whole or in part because of damage to the owner or a separate contractor.

C) **Analysis**

The painting subcontractor failed to adhere to the approved project schedule as required by Section 9.2 of the A401. Specifically, the painter's slow production caused a two-month delay in the completion of the painting work of the hotel rooms. Attached is the baseline schedule that the painter agreed to and the as-built schedule that evidences the painter's delays. It is worth noting that the as-built schedule confirms the contractor turned over the hotel rooms to the painting subcontractor early, yet the painter still could not meet the schedule. Refer to the delay section of this report for a formal schedule analysis.

The contractor complied with Section 3.3.5 of the A401, which requires the contractor to promptly notify the subcontractor of any fault or defect in the subcontractor's work or nonconformity with the subcontract document, as it formally notified the painter its delinquent progress on a weekly basis—a copy of these notifications is attached herein.

Note that Section 4.2.2 of the A401 requires the subcontractor to cooperate with the contractor in scheduling and performing the subcontractor's work to avoid conflict, delay in, or interference with the work of the contractor, other subcontractors, the owner, or separate contractors. The approved schedule was critical to the owner's FF&E contractor, as its work is a successor activity to the painting work in each hotel room. Because of the painter's scheduling delays, this caused the FF&E contractor to be on site for two additional months, as outlined in the owner's claim to the contractor (see attached). As a result, the owner paid the FF&E contractor two months of extended general conditions.

Because the owner backcharged the FF&E contractor's extended general condition claim against the contractor, the contractor hereby backcharges this claim against the subcontractor as it is the sole party that is responsible for this delay, as allowed per Section 11.1.7.2 of the A401, which notes such withholdings are permitted as provided for in Section 9.5.1 of the A201, which permits withholdings due to the damage to the owner or a separate contractor.

D) **Conclusion**

The painting subcontractor's delinquent work caused delays to the owner's FF&E contractor. As a result, the owner paid the FF&E contractor's claim for two months of extended general conditions. Because the owner backcharged the contractor for this amount, the contractor hereby backcharges the subcontractor for this same amount in accordance with the terms of the A401 agreement.

## F.  Change Order Negotiation Issues

Standard contract forms provide the contractor with a remedy in the event that the contractor and a subcontractor reach an impasse during change order negotiations. In such instances, the contractor can direct the subcontractor to perform the work while the cost and time impacts are later worked out. In the event an agreement is not reached, the contractor can trigger dispute resolution.

**Hypothetical Subcontractor Change Order Negotiation Issue:**

The owner retained the contractor under an AIA A101 agreement (design-bid-build with a stipulated sum) that incorporates the AIA

A201 general conditions to construct a four-story multi-family apartment project. Shortly after the start of the work the owner and the architect decided to increase the quantity of metal panel siding versus the fiber cement siding. The architect believes this change should be a nominal cost increase of $3 per square foot while the contractor and subcontractor feel it is an increase of $15 per square foot. As a result of the disagreement, the negotiation stalled and the owner and the architect elected to issue the contractor a construction change directive (CCD) that directed the contractor to proceed with the work and noted the cost increase would be determined in a manner agreed upon by the parties. The siding subcontractor indicated that it refused to proceed with this work; however, after the contractor's receipt of the CCD, the contractor directed the siding subcontractor to proceed with the work per the A401 agreement.

**Sample Entitlement Outline:**

A) **Conclusion**

The owner issued the contractor a change order to reconfigure the siding on the project such that the quantity of metal siding work increases and the quantity of fiber cement siding work decreases. Although the contractor agrees the unit price differential between the products is $15/SF, the architect believes it is only $3/SF, so the owner and architect issued a CCD that forces the contractor to proceed with the work. Accordingly, the contractor hereby directs the subcontractor to proceed with this modification per the terms of the A401. Please prepare a claim for the cost and time impacts of this change and the contractor will incorporate it into its claim to the owner regarding this issue.

B) **Rule**

Section 5.1 of the A401 notes that the owner may make changes in the work by issuing modifications to the prime contract. Upon receipt of an owner modification that affects the subcontractor's work, the contractor is obligated to promptly notify the subcontractor of such modification. Section 5.2 enables the contractor, via written order to the subcontractor, to make changes in the subcontractor's work, including changes required by modifications to the prime contract.

Section 5.3 of the A401 requires the subcontractor to make all claims promptly to the contractor for additional cost and additional time. For claims that will affect or become a part of a contractor's claim under the prime contract, the subcontractor shall make such claims within the specified time and manner such that the contractor can meet the requirements of the prime contract.

Article 1 of the A401 incorporates the A201 general conditions into the subcontract documents. Article 2 of the A401 notes that the A201 general conditions should be interpreted such that the owner is the contractor, and the subcontractor is the contractor.

Per Section 7.3.2 of the A201, the owner and the architect can issue a CCD to the contractor in the absence of total agreement on the terms of a change order. Section 7.3.6 of the A201 notes that upon receipt of a CCD, the contractor shall promptly proceed with the change in the work involved and advise the architect of the contractor's agreement or disagreement with the method, if any, provided in the CCD for determining the proposed adjustment in the contract sum or contract time.

Section 7.3.5 allows the contractor to trigger a dispute resolution under Article 15 if the contractor disagrees with the CCD's position in time. Per Section 7.3.9, the contractor can trigger a dispute resolution in the event that the contractor disagrees with the architect's interim determination of the change in cost related to the CCD.

C) **Analysis**

The owner and the architect issued the contractor a CCD that modifies the configuration of the siding work at the project. A copy of this CCD is attached. In a nutshell, the CCD directs the contractor to increase the quantity of metal siding and decrease the quantity of fiber cement siding on the building veneer. Per Section 5.2 of the A401, this serves as the contractor's written direction to the siding subcontractor to proceed with this CCD modification. Note that the subcontractor previously indicated that it will not proceed with this revision; however, that is not an allowable action under the subcontract.

Both the contractor and subcontractor disagree with the architect's position regarding the cost impacts associated with this CCD. The contractor has already noticed the owner, the architect, and the IDM of a claim that will be submitted regarding this issue. The contractor plans to request an initial decision from the IDM pursuant to Article 15 of the A201. To expedite this claim submission and per Section 5.3 of the A401, the contractor requests that the subcontractor promptly submit its claim for additional costs related to the CCD. The subcontractor's claim will become a part of the contractor's claim to the owner under the prime contract.

If the IDM rejects this claim, the contractor will file for mediation immediately thereafter and if no decision is reached at mediation, the contractor will file for arbitration, all in accordance with Article 15 of the A201.

D) **Conclusion**

Per Section 5.2 of the A401, the contractor directs the subcontractor to proceed with the modification outlined in the attached CCD. The contractor further requests that the subcontractor promptly prepare a claim for the additional cost impacts associated with the CCD work and submit it to the contractor so it can be incorporated into the contractor's claim to the owner per Article 15 of the A201. The contractor will keep the subcontractor advised of the status of this claim as it works through the dispute resolution process.

## VI.  Typical Designer Claims Against Owners

The most common types of claims that designers assert against owners relate to claims for additional services due to owner changes in the program, contractor delays, or contractor maladministration. The following examples assume that the claimant is the architect and the owner-architect agreement is an AIA B101 form that references the A201 general conditions.

### A.  Designer Claim Against Owner for Additional Services Due to Owner's Change in Work Program

Standard owner-architect contracts set forth the owner's program for the project that often includes anticipated physical characteristics such as size, location, dimensions, geotechnical conditions, site boundaries, etc. In addition, most owner-architect contracts will identify the owner's budget for the cost of the work. Architects base their design upon these criteria. If an architect prepares a design that meets these criteria, and then the owner changes the work program that necessitates a redesign, architects often seek reimbursement under the additional services provision on the contract form.

**Hypothetical Designer Claim for Additional Services Due to a Program Change by the Owner:** The owner retained the architect under a B101 agreement that defines the program of work as a two-story office building with a total square footage of 90,000. The architect prepared schematic and design development plans that met this program and the original budgets for construction fit within the anticipated cost of work. Thereafter, the owner decided to go with a three-story building with the same overall square footage of 90,000. Because of this program change, the architect seeks additional design fees under the additional services provision of the contract.

**Sample Entitlement Outline:**

A) **Conclusion**

The owner changed the program for the project by changing the building from two stories to three stories after the architect completed its work relating to the schematic and design development phases. The architect seeks additional compensation under the additional services provision of B101 agreement based on this material programmatic revision.

B) **Rule**

Sections 1.1.1, 1.1.2, and 1.1.3 of the B101 agreement between the owner and architect outline the owner's program for the project. This information includes the type of project, pertinent information about the project's physical characteristics, and the owner's budget for the cost of the work.

Section 4.2.1 requires the architect to promptly notify the owner of the need for additional design services, along with a detailed explanation giving rise for the need. The architect shall not proceed with the additional services until the architect receives the owner's written authorization.

Section 4.2.1.1 notes that additional services may arise if the owner makes a material change in the project including size, quality, complexity, etc. Section 11.3 provides the compensation structure for additional services rendered by the architect.

C) **Analysis**

The B101 agreement clearly sets out the owner's program for the project. Specifically, Sections 1.1.1–1.1.3 outline the owner's program as a two-story, 90,000 square foot, office building for a budget of $32M. The architect prepared schematic design and design development documents that met this program and based on initial discussions with local contractors the design fits with the owner's budget.

The owner recently informed the architect via email that it wishes to revise the program to a three-story building with the same overall square footage. A copy of this email is attached herein. Per Section 4.2.1 of the B101, the architect hereby notifies the owner of the need for additional design services based on this programmatic revision, which will include significant changes to nearly all of the design development drawings as all elevations, plans, and many details will change. Per Section 4.2.1, the owner shall provide the architect with written authorization to proceed before the architect shall perform such additional design services.

Note that Section 11.3 provides the compensation structure for additional services rendered by the architect.

D) **Conclusion**

As required by the B101 agreement, the architect notifies the owner of the need for additional services triggered by the recent programmatic change to the project. Please advise provide written authorization to the architect if you wish to proceed with this change.

## B. Designer Claim Against Owner for Additional Services Due to Contractor Delays

Contractor delays are common on construction projects and these delays often cause extended contract administration for design professionals. Standard owner-designer contracts contemplate extended services to cover project delays, so preparing entitlement narratives for this issue is quite simple. Owners often look to backcharge costs related to extended professional services to the contractor that caused the delay.

**Hypothetical Designer Claim for Additional Services Due to Contractor Delays:** The owner retained the architect under a B101 agreement to provide design and contract administrative services on a commercial office project. Upon completion of the construction documents, the architect put the project out for bid and the owner entered into an AIA A101 agreement (design-bid-build with a stipulated sum) with the lowest qualified bidder. The A101 agreement notes that the contractor would complete the project within 24 months. The contractor was unable to meet this schedule for contractor-caused reasons and, ultimately, the project took 30 months to complete. After month 25, the architect notified the owner of the need for additional services due to the contractor's delays.

**Sample Entitlement Outline:**

A) **Conclusion**

The contractor-caused delays will likely force the architect to provide approximately six months of extended contract administration services on the project. The architect seeks additional compensation under the additional services provision of B101 agreement based on this delay.

B) **Rule**

Section 1.1.4 of the B101 notes the owner's substantial completion date milestone for the project. The A101 owner-contractor agreement lists the substantial completion date in Section 3.3.1.

Section 4.2.1 requires the architect to promptly notify the owner of the need for additional design services, along with a detailed explanation giving rise for the need. The architect shall not proceed with the additional services until the architect receives the owner's written authorization.

Section 4.2.1.4 notes that additional services may arise due to the contractor's failure to perform. Section 11.3 provides the compensation structure for additional services rendered by the architect.

C) **Analysis**

Section 1.1.4 of the B101 lists the owner's substantial completion date milestone for the project, which matches the contractor's substantial completion date for the project. (See Section 3.3.1 of the A101 agreement.) The contractor is now one month beyond this contractual substantial completion date and its most recent project schedule anticipates that the contractor will achieve substantial completion approximately six months beyond the substantial completion date. Below are screenshots of the noted contract terms and the recent schedule update.

Pursuant to Sections 4.2.1 and 4.2.1.4 of the B101, the architect hereby notifies the owner of the need for additional services due to the noted contractor delays. The additional services relate to contract administration per the A201 general conditions, including site visits to inspect the work, OAC meeting participation, submittal and RFI reviews, payment administration, and closeout administration.

Per Section 4.2.1, the architect shall await the owner's written authorization before proceeding with such additional services. Note that Section 11.3 provides the architect's rates for additional contract administration work.

D) **Conclusion**

As required by the B101 agreement, the architect notifies the owner of the need for additional services triggered by the contractor's performance issues. The architect seeks the owner's written authorization to proceed with the additional contract administration on the project.

## C.  Designer Claim Against Owner for Additional Services Due to Contractor Maladministration

When contractors maladminister a contract by issuing needless RFIs or issuing late or incomplete submittals, owner-contractor agreements typically note that the cost of professional services related to the review of these inappropriate submissions is backchargeable. Standard owner-architect contracts also deal with this scenario and allow the architect to request additional services accordingly.

**Hypothetical  Designer  Claim for Additional Services Due to Contractor Maladministration:** The owner retained the architect under a B101 agreement to provide design and contract administrative services on a high-rise condominium project. Upon completion of the construction documents, the architect put the project out for bid and the owner and the successful contractor entered into an AIA A101 agreement (design-bid-build with a stipulated sum). During the construction phase of the project, the contractor maladministered the project by not adhering to the submittal schedule, issuing numerous incomplete submittal packages that had to be resubmitted several times, and issued many superfluous RFIs. The architect seeks to recover the cost related to the time devoted to this unanticipated work.

**Sample Entitlement Outline:**

A) **Conclusion**

The contractor has maladministered the submittal and RFI process on the project, which caused the architect to incur an inordinate and unanticipated amount of time reviewing many superfluous submissions. Accordingly, the architect seeks to recover the additional costs the architect incurred as a result of the contractor's negligent submittal and RFI administration.

B) **Rule**

Section 4.2.2 of the B101 notes that the architect shall provide additional services in order to avoid delay of the construction phase of the project that relate to issues that include, but are not limited to, review of incomplete and out-of-sequence submittals and unnecessary RFIs.

When this happens, Section 4.2.2 requires the architect to notice the owner with reasonable promptness and explain the facts and circumstances giving rise to the need. If, upon receipt of the architect's notice, the owner decides that no further additional services related to such contractor maladministration are authorized, the owner shall advise the architect accordingly and compensate the architect for the additional services the architect provided prior to its notice to the owner.

Section 11.3 provides the compensation structure for additional services rendered by the architect.

C) **Analysis**

Per Section 4.2.2 of the B101, the architect has provided additional services related to the contractor's many incomplete and out-of-sequence submittals and unnecessary RFIs. A list of these submittals and RFIs is attached below, along with a description as to why they were out of sequence, incomplete, or not necessary. The architect has performed these additional services to avoid delay of the construction phase of the project. This letter serves as notice to the owner of these additional services and hereby requests it to authorize continued additional services related to this issue, as necessary. Note that the architect has billed for these additional services to date per the rates set forth in Section 11.3 of the B101.

D) **Conclusion**

As required by the B101 agreement, the architect notifies the owner of its additional services performed to date regarding the contractor's maladministration of the submittal and RFI process. The architect seeks payment for these additional services and requests written authorization from the owner to either continue or halt further review of these deficient contractor items.

## VII. Typical Owner Claims Against Designers

Typical owner delays against designers relate to design errors or omissions, designer maladministration of contract administration duties, or design delays for which the design is responsible.

### A. Claim for Design Error or Design Omission

Design issues are tricky because no set of construction documents is perfect and design-related change orders are commonplace on construction projects so the issue is, when should an architect be responsible to the owner for a design error or omission? According to an article published by the AIA, "it depends on the circumstances, on the contract, and, perhaps most of all, on whether or not the owner and architect started out with shared expectations on the seemingly arcane subject of standard of care."[1] Section 2.2 of the AIA B101 defines the standard of care by requiring the architect to "perform its services consistent with the professional skill and care ordinarily provided by architects practicing in the same or similar locality under the same or similar circumstances. The Architect shall perform its services as expeditiously as is consistent with such professional skill and care and the orderly progress of the project."

While the AIA's "reasonable and similarly situated architect" guide makes sense, a rule of thumb would certainly help discern when a designer's mistakes go from expected to unexpected, or foreseeable to unforeseeable. My take on this subject is based on a review of the B101 document. Article 6 of the B101, Cost of Work, defines budget items that the architect can include within its estimate to construct all elements of the subject project that will be designed or specified by the architect. Section 6.3 allows the architect to include contingencies within the cost of work for design, bidding, and price escalation. Thus, the cost of work should include contingent funds for design issues. Based on experience, a typical design contingency for most projects should initially fall within the 3–5% range of the contractor's initial contract sum. Per RSMeans, the most widely used construction cost database used in North America, the typical contingency amount on final design drawings is 3%. I would qualify this as relating to projects where the contractor's initial contract sum is greater than one million dollars. As work progresses, this contingent consideration should be reduced if the work progresses without the identification of any design errors or omissions. Accordingly, as a rule of thumb, costs related to design errors or omissions that materially extend beyond this threshold contingency amount might be considered a designer responsibility. As noted by AIA, the circumstances, the contract, and the defined standard of care definition should be weighed as well.

**Hypothetical Owner Claim for a Design Error:** The owner retained the architect under a B101 agreement to provide design and contract administrative services on a higher education project. The architect specified non-fire-rated doors to the classrooms. After installation of the doors and hardware the fire marshal conducted an inspection of the work and rejected the contractor's request for a temporary certificate of occupancy because the classroom doors were not fire rated. The architect's estimate of the cost of work for the project included a 5% contingency on top of the contractor's bid to account for issues such as design errors and omissions. As the project neared substantial completion, this contingency has been exhausted, and then some, to address the costs associated with design issues that the contractor pointed out during the course of construction. The owner now seeks recovery from the architect because of this design issue. Note that if this claim goes to binding dispute resolution, the owner will likely need to retain an architectural standard of care expert to render an opinion on the liable party for this issue.

**Sample Entitlement Outline:**

  A) **Conclusion**

    The owner seeks reimbursement of the cost of the change order issued to the contractor to switch the currently installed classroom doors that match the requirements of the contract documents, to the fire-rated doors that are required by code, as recently pointed out by the fire marshal. To date, the owner has exhausted its contingency for design changes on the project.

B) **Rule**

Per Section 2.2 of the B101, the architect shall perform its services consistent with the professional skill and care ordinarily provided by architects practicing in the same or similar locality under the same or similar circumstances. The architect shall also perform its services as expeditiously as is consistent with such professional skill and care and the orderly progress of the project.

Section 2.5.6 of the B101 requires the architect to furnish professional liability insurance that covers the negligent acts, errors and omissions in the performance of professional services. Section 3.4.2 requires the architect to incorporate the design requirements of governmental authorities having jurisdiction over the project into the construction documents.

Per Section 6.1 of the B101, the architect's estimate of the "cost of the work" shall be the total cost to the owner to construct all elements of the project designed or specified by the architect and shall include contractors' general conditions costs, overhead and profit. The cost of the work does not include the compensation of the architect; the costs of the land, rights-of-way, financing, or contingencies for changes in the work; or other costs that are the responsibility of the owner.

Per Section 8.1.1 of the B101, the owner and the architect shall commence all claims and causes of action against the other and arising out of or related to the B101, whether in contract, tort, or otherwise, in accordance with the requirements of the binding dispute resolution method selected in the B101 within the period specified by applicable law, but in any case not more than 10 years after the date of substantial completion of the work.

Per Section 11.10.2.2, the owner shall not withhold amounts from the architect's compensation to impose a penalty or liquidated damages on the architect, or to offset sums requested by or paid to contractors for the cost of changes in the work, unless the architect agrees or has been found liable for the amounts in a binding dispute resolution proceeding.

In terms of dispute resolution, Section 8.2 mandates mediation as a condition precedent to binding dispute resolution and if unsuccessful, Section 8.2.4 lists the types of binding dispute resolution that parties can elect. For this project, the parties selected arbitration, which is further defined in Section 8.3.

C) **Analysis**

Section 3.4.2 of the B101 requires the architect to incorporate the design requirements of governmental authorities having jurisdiction over the project into the construction documents. The architect failed to do this in terms of stipulating fire-rated doors at the classrooms at the project. Attached are the contract documents that relate to these doors, and as shown, the specified classroom doors do not have a two-hour fire rating, which is in violation of local code, which is also attached herein.

This error required a modification to the design after the doors were furnished and installed by the contractor. The fire marshal recently rejected these doors and requires the two-hour doors to be installed per local code. A copy of the fire marshal's inspection notes is attached herein. The owner understands that no design for a project of this magnitude will be perfect, and that is why the owner included a 3% contingency amount for design changes based on the cost of the work estimated by the architect per Section 6.1 of the B101.

To date, this contingency amount has been exhausted and then some due to design changes caused by design errors and omissions. At this point the costs related to design issues amount to over 10% of the cost of the work. A breakdown of these costs to date is listed below. Moreover, a copy of the change order that the architect and owner authorized to the contractor for changing the existing doors to fire-rated doors is included herein.

Per Section 11.10.2.2, the owner is not able to withhold this amount from the architect unless the architect agrees or has been found liable for the amounts in a binding dispute resolution proceeding. Hence, the owner requests the architect agree to allow the owner to backcharge this amount against the architect's remaining contract balance. If the architect does not agree to this backcharge or otherwise agree to reimburse the owner, then the owner will trigger dispute resolution per Article 8 of the

B101, which note mediation and then arbitration, if mediation is unsuccessful.

D) **Conclusion**

The owner seeks agreement from the architect to allow the owner to backcharge the architect, or an agreement from the architect to reimburse the owner, for costs relating to the design error relating to the fire rating of the classroom doors, for which the owner must fund a change order to the contractor to address. To date, the architect's design errors on the project have exceeded the contingency amount that the owner included on top of the estimated cost of the work for the project; thus, this request is reasonable.

## B.  Owner Claim for Designer Maladministration

When designers maladminister contracts by failing to respond promptly to administrative issues, such as RFIs, submittals, payment applications, or inspection requests, it can have a detrimental effect on the schedule and cost of the work. Typically claims of this nature involve a pattern of delinquent behavior on the designer's part, rather than just a few late responses. When a pattern of delinquent behavior like this transpires, it can cause damage to the owner vis-à-vis contractor claims, and in such instances the owner may assert a claim against the designer for maladministration.

<u>**Hypothetical Owner Claim Due to Designer Maladministration:**</u>
The owner retained the architect under a B101 agreement to provide design and contract administrative services on a high-rise condominium project. Upon completion of the construction documents, the architect put the project out for bid and the owner and the successful contractor entered into an AIA A101 agreement (design-bid-build with a stipulated sum). During the construction phase of the project, the architect maladministered its duties by not responding to many RFIs promptly and it did not return submittals per the time limits noted in the approved submittal schedule. These actions caused a delay to the project and caused the owner to pay the contractor extended general conditions. The owner seeks reimbursement from the architect for amounts paid to the contractor.

**Sample Entitlement Outline:**
A) **Conclusion**

The architect maladministered the contract by failing to respond to RFIs promptly and failing to review critical submittals per the durations noted in the approved submittal schedule. As a result, the owner had to fund the contractor's request for extended general conditions. The owner seeks approval from the architect to offset these costs from the architect's contract balance.

B) **Rule**

Per Section 3.6.1.1 of the B101, the architect shall provide administration of the contract between the owner and the contractor as set forth in the B101 and per the A201 general conditions. Section 3.10.2 of the A201 notes that the contractor shall submit a submittal schedule for the architect's approval. Once the architect approves the submittal schedule, the contractor shall issue submittals and the architect shall review submittals per this schedule, as reaffirmed in Section 4.2.7 of the A201. If the architect fails to review the contractor's submittals per the approved schedule, and this impacts the overall schedule for the project, the contractor is entitled to an adjustment in the contract time for this delay.

In addition, Section 3.6.4.4 of the A201 notes that the architect will review and respond to the contractor's requests for information about the contract documents within any time limits agreed upon or otherwise with reasonable promptness and, if appropriate, the architect shall prepare and issue supplemental drawings and specifications to the requests for information.

While "reasonable promptness" is an undefined term in the contract documents, several studies have been conducted by construction professionals to define what reasonable promptness means. According to page 8 of Navigant Consulting's Impact & Control of RFIs on Construction Projects White Paper, the average reply time for RFIs is 10 days, based on a review of approximately 1,400 projects.

Per Section 11.10.2.2, the owner is not able to withhold amounts from the architect's payment application unless the architect agrees or has been found liable for the amounts in a binding dispute resolution proceeding. Hence, if the architect does not agree to reimburse the owner, then the owner will trigger dispute resolution per Article 8 of the B101, which notes mediation and then arbitration, if mediation is unsuccessful.

C) **Analysis**

The B101 agreement notes that the architect will provide contract administrative services per the B101 and the A201. A copy of the B101 and A201 are attached herein.

The architect fell short on its administrative duties related to submittals and RFIs on the project. The contractor submitted and the architect approved the submittal schedule for the project. However, the architect violated Section 4.2.7 of the A201 by not reviewing and responding to many of the contractor's critical submittals per the durations stipulated in the approved submittal schedule. Below is a table of the contractor's submittals compared to the submittal schedule and the architect response duration when compared to the submittal schedule. A review of this information clearly demonstrates that the contractor issued its submittals in a timely manner, but the architect took over twice as long to return submittals than the durations allotted to the architect in the approved submittal schedule. On the late submittal items, the architect made no complaints regarding the adequacy of the contractor's submittal packages, it just took longer than required.

Similarly, the architect failed to timely respond to the contractor's RFIs on the project. Section 3.6.4.4 of the A201 requires the architect to review and respond to the contractor's submittals with reasonable promptness. This did not happen. As shown in the table below, the architect's average response time for RFIs was 45 calendar days, which is not reasonably prompt. This duration falls well beyond the time recommended in the industry study that Navigant conducted to measure average turn times for RFIs in the construction industry. A copy of this study is included herein.

The contractor prepared a forensic schedule analysis that details the impacts of the architect's delinquent administration. The contractor included this in its claim to the owner, the architect, and the IDM, per Article 15 of the A201. The architect has since issued a change order to the contractor, which the owner signed and funded, for the time and cost impacts related to this claim. A copy of this claim and the change order are included herein.

Per Section 11.10.2.2 of the B101, the owner is not able to withhold amounts from the architect's payment application unless the architect agrees or has been found liable for the amounts in a binding dispute resolution proceeding. Hence, the owner requests that the architect agree to a backcharge in the amount of the contractor's change order for the architect's maladministration on the project. If the architect does not agree, the owner will trigger dispute resolution per Article 8 of the B101.

D) **Conclusion**

The owner seeks approval from the architect to offset the architect's unpaid contract balance by the amount of the contractor's change order related to the architect's maladministration of the submittals and RFIs on the project. In the event the architect disagrees with this request, the owner will likely trigger dispute resolution in order to get a binding decision on this issue.

## C.  Owner Claim for Design Delays

Standard owner-designer contracts stipulate the designer's schedule for the various design phases of the project. In design-bid-build projects, the owner relies on the designer to timely generate contract documents because the completed design is the predecessor to the procurement and construction phase of work. If the project delivery for the project contemplates an overlap of the design and construction phase, such as in construction management and design-build forms of delivery, the owner and the contractor rely on the designer to timely complete contract documents in order to avoid impacts to the construction schedule. When designers falter and fail to complete the design of a project per the terms of the owner-designer agreement, and the delay causes a detrimental impact to an owner, this gives rise to a potential claim by the owner against the designer.

**Hypothetical Owner Claim for Design Delays:** The owner retained the architect under a B101 agreement to provide design and contract administrative services on a mixed-use project. The owner plans to retain a contractor under an A101 agreement (design-bid-build with a stipulated sum). The B101 agreement notes that the design phase should be complete by February, which gives the designer eight months to complete all three phases of the design (schematic, design development, and construction documents). If the architect completes its design by February, the contractor will likely be able to start work several months thereafter and be dried in by November, to avoid excessive winter protection costs. As it turns out, the architect did not complete the design until August, or six months late, which caused an increase in the cost of the work due to winter protection costs. The parties agree the delay is the responsibility of the architect. The owner seeks to withhold the architect's fees related to the construction documents phase of the design, which represents approximately half of the winter protection costs that it will incur due to the late design.

**Sample Entitlement Outline:**

A) **Conclusion**

The architect delayed the design phase of the project by six months as it informed the owner that it was too busy with other projects to meet the milestones set forth in the B101 agreement. As a result, the owner must now pay the contractor unanticipated winter protection costs because it will not be able to dry in the building before the winter months. The owner seeks agreement from the architect to withhold the design fees related to the construction documents phase of the work in order to partially offset the contractor's costs related to winter protection.

B) **Rule**

Section 1.1.4.1 of the B101 notes milestone dates for the various phases of the design and construction for the project. Per Section 11.10.2.2 of the B101, the owner is not able to withhold amounts from the architect's payment application unless the architect agrees or has been found liable for the amounts in a binding dispute resolution proceeding.

C) **Analysis**

Section 1.1.4.1 of the B101 notes a milestone for the completion of 100% construction documents to be in February. A copy of the B101 is attached herein. Unfortunately, the architect did not meet this milestone date and ultimately delivered the completed design in August, or six months later than planned. The architect issued several communications to the owner during that delay period, which are annexed to this letter, that confirm the design delay is solely related to the architect being too busy with other projects to meet the deadlines on this project.

The reason that the owner and architect defined a February deadline was to allow the successful contractor to dry in the building by November, which would save the owner money because winter protection costs would be greatly mitigated. That is also why the architect included minimal winter protection costs in its estimate of the cost of the work (see attached), that is prepared per Article 6 of the B101.

The architect worked diligently on the schematic and design development phases of the work, but the construction documents phase of the design took many months longer than anticipated. A detailed review of the design milestones and the architect's submissions can be found in the delay section of this letter. In addition, a review of the contractor's proposed winter protection costs can be found in the damages section of this letter.

Per Section 11.10.2.2 of the B101, the owner is not able to withhold amounts from the architect's payment application unless the architect agrees or has been found liable for the amounts in a binding dispute resolution proceeding. Hence, the owner requests that the architect agree to a backcharge in the amount of the architect's fees related to the construction documents phase of the work, which is a fraction of the cost of the proposed winter protection costs associated with the contractor not being able to be dried in by the winter months.

D) **Conclusion**

The architect admittedly delayed the construction documents phase of the design by six months, and this delay will cause the owner to incur significant unanticipated winter protection costs because the contractor will not be able to dry in the building before the winter months. The owner seeks agreement from the architect to withhold design fees related to the construction documents phase of the design to help offset the anticipated winter protection costs related to the architect's delay.

## VIII.  Summary

Many claimants skip or gloss over proving entitlement for claims and jump directly into delay and damage discussions. However, if entitlement is not established, a claimant cannot recover time or money so it is imperative that time be spent in preparing claims in order to walk the respondent through the reasons why the claimant is entitled to recover time or money related to the claim. In most instances, entitlement is a detailed review of what the obligations and expectations are per contractor against actual conditions, and then drawing a comparison between the two and proving how the respondent is responsible for the impacts caused by this difference in condition or expectation.

For outline purposes, I recommend that claimants frame entitlement arguments using the CRAC method. While headings don't need to be titled Conclusion, Rule, Analysis, and Conclusion, the flow of information to the reader makes sense when it is organized in this manner. Again, claims are not mystery novels and respondents want to know up front what claimants are requesting, rather than waiting until the end of the narrative to understand why claimants believe they are entitled to additional time and/or money.

## Note

1  See https://info.aia.org/blast_images/kc/pmkc_12_winter_digest.html; https://www.aia.org/best-practices/6336320-standard-of-care-confronting-the-errors-an

# 6

## Step 5: Calculate Delay

Once entitlement is established for an impact issue, the next step in the claims process is to evaluate whether the impact caused a delay to the substantial completion date of the project through a forensic schedule analysis. It is important for the claimant to review the contract requirements related to delay claims to determine what type of delay analysis is required, if any. While many variants of the methodologies discussed in this chapter exist, the most common forensic scheduling techniques are classified as one of the following:

1) as-planned v. as-built analysis;
2) windows analysis;
3) time impact analysis;
4) collapsed as-built analysis.

Depending on the nature of the dispute and the information available to the claimant, certain methodologies are more appropriate than others.

It is also important for the claimant to determine if delay-related damages are recoverable under the contract. Claimants can only recover delay damages for *compensable* delays. When the claimant concurrently delays the work, or if the contract does not allow the claimant to recover delays damages for the claimed type of impact, the delay is defined as *excusable* but non-compensable. In such instances, the claimant is due an adjustment to the contract time as its sole remedy for the excusable portion of the impact. Claimants often assert that concurrent delays are a result of purposeful pacing of the non-critical work and this, in certain instances, converts an excusable delay into a compensable delay.

For this chapter, the claimant will be an at-risk general contractor and the respondent will be the project owner that has a standard contract agreement with the general contractor. Note that the methodologies described herein would equally apply to claimants such as subcontractors, vendors, and other parties involved in a construction project.

## I. Contract Requirements for Time Extension Requests

Most standard contract forms do not prescribe a specific technique for delay analysis. To the contrary, proprietary contracts, particularly those used on federal projects, often mandate a specific forensic scheduling methodology for time extension requests.

A) **AIA A201 General Conditions:** Does not prescribe a type of forensic scheduling analysis to support time extension requests.
B) **ConsensusDocs 200:** Does not prescribe a type of forensic schedule analysis to support time extension requests.
C) **EJCDC C700:** Does not prescribe a type of forensic schedule analysis to support time extension requests. However, Section 4.05E does note that the owner or the engineer may require the contractor to furnish a revised progress schedule that identifies all activities affected by the impact and an explanation of the effect of the impact on the critical path.

## II. Scheduling Overview

A Critical Path Method (CPM) schedule takes information about tasks, task durations, and task dependencies and calculates the minimum time required to complete a project. Contractors utilize software such as Primavera, Microsoft Project, and other scheduling software to prepare CPM schedules. A contractor's baseline schedule is the contractor's initial schedule for the project that is submitted for owner approval. The as-planned schedule is typically done before the start of physical work.

In developing the schedule, the contractor creates activities for the tasks required to complete the project and a duration to each. The tasks are then linked via activity logic based on their interdependencies. These interdependencies may be based on physical constraints or preferential relationships such as crew logic. Based on the duration and logic assigned, the CPM software calculates the earliest and latest date each activity can start and finish. The difference between an activity's early and late dates is the amount of total float available to the activity. Total float is simply the number of days that an activity can fall behind before it impacts the completion date for a project. The activity sequence with the least amount of total float is the critical path, which is the defining characteristic of a CPM schedule.

The critical path notes the activities that need to complete on time in order to maintain the forecasted completion date of a project. For instance, on a single-family house project, the critical path of the as-planned schedule would typically run through excavation, foundation, framing, roof dry-in, drywall, and then interior finishes. Thus, a delay to the excavation work would impact critical successor activities and cause a delay to the project completion date, absent acceleration or other recovery efforts.

As work progresses, contractors typically update project CPM schedules at intervals required by contract, which is often on a monthly basis. Because the baseline schedule is an educated estimate on how work will likely progress, it is inevitable that the actual schedule dates will differ from as-planned dates. In addition, the schedule may evolve over time to reflect changes to the project such as if the owner authorizes change orders that modify the contractor's scope of work. Additional work activities related to the change orders should be inserted into the schedule to identify any impact such changes have on the completion of the project, and to ensure the schedule models all project requirements. Also, if the owner causes a detrimental impact to the contractor's work, activities modeling the impact can be added into the schedule to monitor its effect on the project.

## III.   Types of Delays

Delays are classified as either excusable or non-excusable. Non-excusable delays arise from the actions or inactions of the claimant and therefore do not entitle the claimant to time or delay damages. Non-excusable delays may also entitle the respondent to assert actual delay damages or liquidated damages against the claimant, depending on the contract agreement. Conversely, excusable delays are those which are beyond the control of claimant. At a minimum, excusable delays entitle the claimant to a time extension. Whether the claimant is also entitled to recover certain delay-related damages depends on whether the excusable delay can be further classified as a compensable delay.

Excusable delays are either compensable or non-compensable. To be compensable, the delay must typically meet two conditions. First, the delay must be caused by the actions or inactions of the owner or a party for which it is responsible. Common owner-responsible delays include those caused by design errors, contract modifications, and administrative delays. Second, the respondent delay must not be concurrent with a claimant-caused or other non-excusable delay. If the excusable delay meets these conditions, it is generally considered a compensable delay and entitles the claimant to both time and certain delay-related damages.

If the delay is excusable to the claimant but beyond the control of the respondent, then the delay is non-compensable, and the claimant is typically limited to a time extension as a remedy. Examples of common excusable non-compensable delays include those caused by unusually severe weather or acts of God.

## IV. Concurrent Delays

Concurrent delays occur when two or more parties cause delays to the project during all or a portion of the same time period. For instance, a concurrent delay exists if the contractor and the owner cause delays to certain work on the project and if either delay is removed from the schedule, there remains an impact to the completion date of the project. Here, the claimant maintains a compensable delay claim for the non-concurrent portion of the impact. The claimant also maintains an excusable delay claim to the extent of the concurrency, unless the claimant can prove the concurrency is a result of an election to pace the work due to the respondent's delay, which is discussed in the next section of this chapter.

It is important to note that most contract forms do not contemplate scenarios where concurrent delays exists. Moreover, some experts argue that a concurrent delay only exists if it is a "pure" concurrent delay—when both parties' delays impact the critical path at the exact same time. In a pure concurrency scenario, the delay events share start and finish dates, either of which occurring by itself will cause the same delay to project. An example pure concurrent delay scenario is illustrated in Figure 6.1.

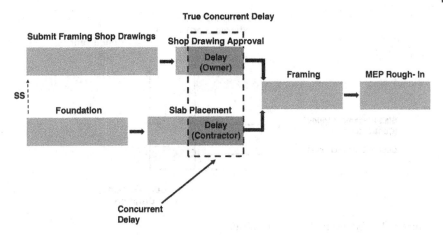

**Figure 6.1**  Example of true concurrent delay.

Pure concurrency scenarios are rare. It is much more common for concurrent delays to be viewed from a functional sense where the delays do not necessarily overlap but occur during the same time period, and either of which absent the other, would independently delay the critical path. For example, assume a contractor asserts a delay caused by an owner design issue impacted the project's substantial completion date by 90 days and the contractor therefore claims it is entitled to a compensable time extension of 90 days. The owner, however, asserts that a contractor performance issue would have caused a 70-day delay to the critical path but-for the owner design issue. In this scenario, the contractor's remedy would typically be a 70-day excusable non-compensable time extension and a 20-day compensable time extension. In addition, the owner would not be able to assert any actual delay damages or liquidated damages for the entire 90-day excusable period (Figure 6.2).

It is important to note that the concept of concurrent delays is not explicitly defined, and no consensus exists within the industry as to how handle concurrent impacts. Contracts rarely provide clarity on the topic which in turn allows for flexibility among the parties as to what is considered a concurrent delay. Parties to a construction project should work to establish guidelines on how to address concurrent impacts during the early stages of a project whenever possible.

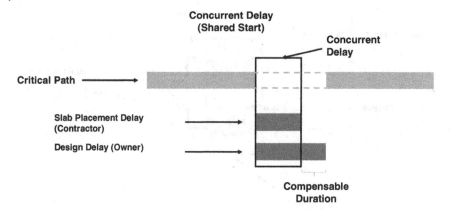

**Figure 6.2** Example of concurrent delay.

## V. Pacing Delays

Claimants often challenge a respondent's concurrent delay argument by proving the concurrent delay was a result of purposeful "pacing." When critical path activities are delayed, the total float of non-critical activities will increase due to the additional time available to complete the work. Pacing occurs when the contractor takes advantage of the additional float created by an owner-caused critical path delay and intentionally slows down the progress of certain near-critical work. This defense is controversial because pacing is nearly always an undefined term in contracts and owners often take the position that contractors are obligated to maintain the schedule despite owner-caused delays as contracts often note that "time is of the essence" for contractors. Claimants also typically fail to document pacing decisions which results in the respondent often perceiving the pacing argument as nothing more than an after-the-fact excuse for a non-excusable delay.

For instance, assume a project's critical path runs through framing, then roof dry-in, then interior drywall, and finally finishes. A near critical path runs through rough mechanical, electrical, and plumbing work in the interior of framed walls, which is also a predecessor to the critical interior drywall. Further assume that the contractor's baseline schedule calls for the roof to be dried in by November 1, and interior drywall was to commence immediately thereafter. In addition, the baseline schedule notes that rough mechanical-electrical-plumbing (MEP) work is to be done by mid-October, two weeks before the scheduled roof dry-in date. Hence, the rough MEP work has two weeks of float. During the course of construction, the contractor identified an owner design issue with the roof, and that caused significant redesign work, which ultimately caused a one-month delay to the roof dry-in work. Thus, the updated schedule noted that contractor was to complete the roof dry-in work on December 1, or one month after the originally planned dry-in of November 1.

The contractor's MEP subcontractors originally planned on increasing their manpower in September to meet their mid-October deadline. However, due to the owner's design delay which pushed out the roof dry-in to December 1, the MEP activities gained an additional one (1) month of float. As a result, the contractor did not mandate that its MEP subcontractors increase their manpower to meet the initial as-planned date as this would have caused a "hurry up and wait" situation. Accordingly, the MEP subcontractors paced their work and completed the rough MEP work in mid-November, or one month later than originally scheduled and two weeks before the actual roof dry-in work. The contractor later submitted a claim due to the owner's design issue asserting a one-month compensable delay. The owner rebutted the claim, noting the contractor's two-week concurrent delay associated with the late finish of the rough MEP work relative to the as-planned schedule. The contractor countered by noting that the late completion of the rough MEP work was resultant from a purposeful and prudent pacing decisions, and the contractor should not be penalized for this election.

Courts have considered and addressed this issue. In general, courts have held that if the contractor can establish that it relaxed its performance of certain portions of work due to owner-caused delays, and this purposeful relaxation did not impact the completion date of a project, the relaxation should not be seen as a concurrent delay that penalizes the contractor. However, there are practicable problems of pacing that go beyond the lack of definition as it runs against "time is of the essence" clauses of most contracts. Also, owners can often raise a lack of notice argument to counter a pacing delay defense, particularly because contractors rarely issue notice for pacing decisions.

## VI. Review of Forensic Scheduling Methodologies

The selection of a forensic schedule analysis depends on the contract requirements, available schedule data, reliability of the available information, time constraints, and other circumstances particular to the given situation. The construction industry has four commonly used methodologies for forensic schedule analysis to prove delays on a construction project. These methodologies include:

1) as-planned v. as-built analysis;
2) windows analysis;
3) time impact analysis;
4) collapsed as-built analysis.

This section will discuss the key principles of each of the above methodologies. The reader should note that variations of each method are recognized by the industry as detailed in other trade publications.[1]

Forensic scheduling methods refer to those which are retrospective in nature. Prospective analyses are performed prior to or contemporaneous with the delay event. Hence, a prospective analysis estimates the impact the event will have on the completion date of a project going forward. Of the four commonly used techniques, the time-impact analysis is the only method that can be used prospectively. Retrospective analyses are performed after the delay event has occurred and the impacts are known. Thus, the analyst has the benefit of hindsight when performing a retrospective analysis. All four techniques discussed below can be used in a retrospective analysis.

### A. As-Planned vs. As-Built Analysis (Retrospective, Backward-Looking)

The as-planned versus as-built schedule delay analysis is a retrospective method which involves comparing the as-planned construction schedule against the as-built schedule or a contemporaneous schedule update. An as-planned vs. as-built analysis is typically implemented when a reliable baseline and as-built activity data exist, but interim schedule updates were either not produced or are not reliable to support a delay analysis.

The as-planned versus as-built method simply compares what actually happened to what was supposed to happened. It measures the activities of the baseline or other planned schedule against that of the as-built schedule or contemporaneous schedule updates to identify encountered delays. When using this method, it is important that delays are measured by the differences between the as-built actual dates and the *late* dates in the as-planned schedule as opposed to the early dates. The late dates reflect when available float for the activity is exhausted and thus when the activity becomes critical.

Inadequate project record documentation is the most frequently encountered obstacle to the effective use of certain delay analysis methods. Ideally, the effect of a delay impact on the schedule is measured at the time it occurs. To do so, the project schedule must be properly maintained by regularly updating it to reflect actual start and finish dates, reassessed activity durations, modified activity relationships, etc. Oftentimes, contractors fail to adequately update the project schedule thus precluding the use of some of the more technical delay analysis methodologies. In these cases, the as-planned versus as-built method is an attractive option, as it does not require contemporaneous updates.

One of the greatest strengths of the as-planned versus as-built method is its simplicity. Even in the most straightforward disputes, an overly complex analysis can diminish chances of recovery. If a delay analysis is too complicated to follow, then the evaluator will not trust its findings. The as-planned versus as-built method is very easy to understand and thus appealing to those presenting to an inexperienced audience. With the simplicity of the method comes the added benefit of low-resource requirements. Because the method is less intensive, the time required to prepare the analysis is typically significantly less than that of other more complex methods.

While as-planned versus as-built method is straightforward, requires fewer resources to implement, and overcomes limitations with available documentation, it does have weaknesses that should be considered. For example, the method is not well suited to determining delay causation when a project is complex, spans longer durations, or is built in a manner than significantly deviates from the planned sequence. Additionally, the method is not typically appropriate for projects that include near critical paths that experienced delay, as the method cannot consider shifts in the critical path over time.

Another limitation of the method is its failure to consider concurrent delays. The as-planned versus as-built method will not typically depict these critical considerations. Special attention to documenting concurrent impacts or known pacing events when performing the analysis is critical.

In conclusion, the as-planned versus as-built method is relatively easy to perform and present. However, it might not be suitable for complicated projects or projects built significantly different than planned. This methodology is generally accepted by the courts but is viewed critically when documentation substantiating relied upon as-built activity dates does not exist or other more comprehensive methods were available.

**Example – *As-Planned vs. As-Built Analysis***
The contractor's agreement with the owner requires the construction of the small building project within 120 days, or 4 months.

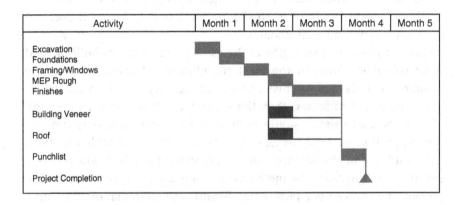

The contractor completed the work within 150 days; thus, the contractor completed the work 30 days late. The contractor argues that the late completion was a result of abnormal weather that occurred during the installation of foundations. As shown below, the baseline schedule is compared to the as-built schedule to reflect the abnormal weather impact. The agreement between the contractor and the owner stipulates that time but no money is the remedy for abnormal weather delays. Accordingly, the contractor is entitled to a 30-day excusable non-compensable time extension.

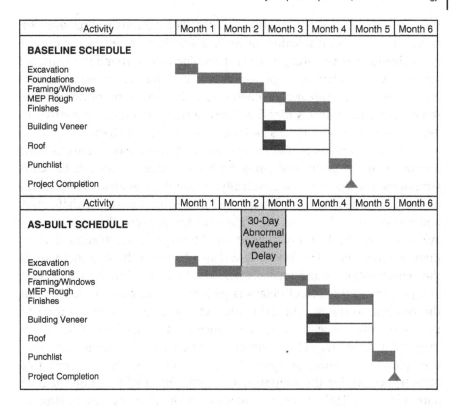

## B. Windows Analysis (Retrospective, Forward-Looking)

A windows analysis, sometimes referred to as a contemporaneous period analysis, is a retrospective technique that involves interim assessment of delays on updated schedules over specific intervals or "windows." Fundamentally, a windows analysis is simply a study of the development of the project's critical path over time. This study is segmented into discrete intervals (windows) that allow for the identification of explicit causes of slippage to the critical path.

In applying this methodology, the overall performance period being analyzed is partitioned using available contemporaneous progress updates with the start and finish dates for each window typically determined by the data dates of two consecutive updates. The analysis begins with the baseline construction schedule or earliest schedule preceding the project delays being analyzed and proceeds chronologically through the available updates. For each window, the as-built schedule for the previous window is assigned as the as-planned schedule for the evaluation of delays over the window. The impacts for all window periods are then summarized to determine the overall impacts and associated responsibilities.

A windows analysis is effectively a series of as-planned versus as-built analysis that are performed using each successive pair of reliable contemporaneous updates. Mechanically, a windows analysis is identical to the as-planned vs. as-built analysis in that it compares activity start and finish dates, durations and relationships between the updates and evaluates the effect of any change on the project's completion date or interim milestones. The key differentiator between the two methods is the segmentation of the analysis periods. In an as-planned vs. as-built analysis, the evaluation of delays is performed over one window—from the baseline to the as-built schedule. As noted above, one criticism of the as-planned vs. as-built is its reliance on the baseline logic, which may differ from the actual project conditions, to determine criticality throughout the entire analysis. The windows analysis addresses this issue by performing the analysis over shorter intervals based on the data dates of the available contemporaneous schedule updates. By relying on interim updates, the as-planned schedule for each window resets to reflect the contemporaneous critical path and therefore considers the dynamic nature of construction and changes to the project's critical path over time.

A windows analysis is the most comprehensive type of retrospective analysis, so it is widely accepted by courts and agencies. Fundamentally, the windows analysis method utilizes contemporaneous schedule updates, in conjunction with as-built data, to quantify the delay impact. For this reason, the successful execution of a windows analysis depends on the availability of reliable baseline schedule information, contemporaneous progress updates, and verifiable as-built data. However, locating the necessary information to perform a windows analysis is often a problem, particularly on smaller projects and certain types of work that do not generally involve sophisticated schedule reporting. Moreover, when drastic logic and/or activity changes occur from schedule to schedule, it makes this type of analysis difficult. Lastly, it is the most time-consuming and thus most expensive type of schedule analysis to perform in most cases.

## Example – *Windows Analysis*

Below is a simplified baseline schedule for a building project. Assume the contractor encountered three delays over the course of construction. A windows analysis is used to determine which party is responsible for the various delays. Red activities are critical, blue activities are non-critical, and green activities represent delays.

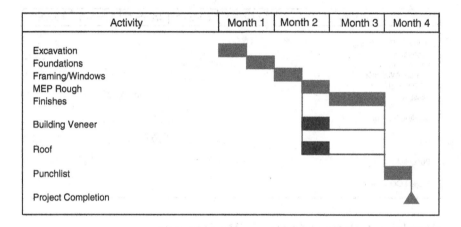

- Excusable Compensable Delays:          0
- Excusable Non-Compensable Delays:      0
- Non-Excusable Delays:                  0
- Overall Delay:                         0

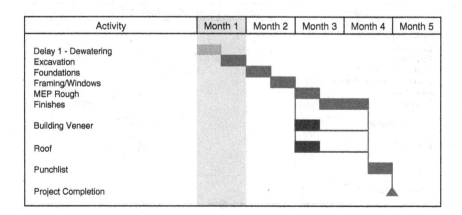

## Window 1 above Month 1

- Excusable Compensable Delays:     15 [differing site condition]
- Excusable Non-Compensable Delays:   0
- Non-Excusable Delays:             0
- Overall Delay:                    15

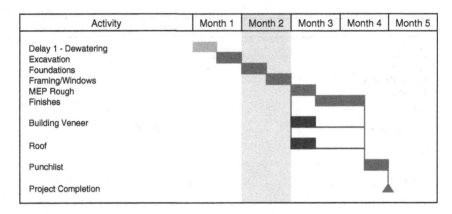

## Window 2 above Month 2 [No Delays]

- Excusable Compensable Delays:     0
- Excusable Non-Compensable Delays:   0
- Non-Excusable Delays:             0
- Overall Delay:                    15

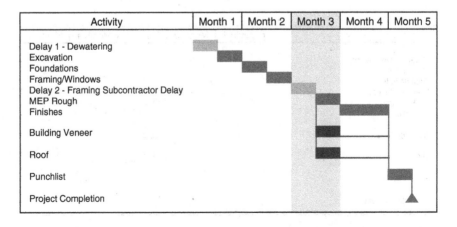

# Window 3 above Month 3

- Excusable Compensable Delays:       0
- Excusable Non-Compensable Delays:   0
- Non-Excusable Delays:               15 [subcontractor delay]
- Overall Delay:                      30

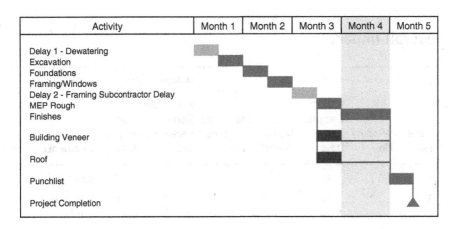

# Window 4 above Month 4 [No Delays]

- Excusable Compensable Delays:       0
- Excusable Non-Compensable Delays:   0
- Non-Excusable Delays:               0
- Overall Delay:                      30

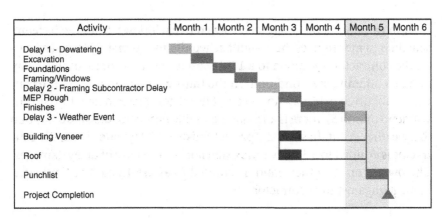

## Window 5 above Month 5

- Excusable Compensable Delays: 0
- Excusable Non-Compensable Delays: 15 [weather delay]
- Non-Excusable Delays: 0
- Overall Delay: 45

## Overall Impact

| | | | | | Delays | | |
|---|---|---|---|---|---|---|---|
| Month No. | Window No. | Project Completion Days | Delay During Period | Excusable Compensable | Excusable Noncompensable | Non Excusable | Comments |
| 0 | 1 | 105 | 0 | 0 | 0 | 0 | Baseline |
| 1 | 2 | 120 | 15 | 15 | 0 | 0 | Differing site condition |
| 2 | 3 | 120 | 0 | 0 | 0 | 0 | |
| 3 | 4 | 135 | 15 | 0 | 0 | 15 | Subcontractor delay |
| 4 | 5 | 135 | 0 | 0 | 0 | 0 | |
| 5 | 6 | 150 | 15 | 0 | 15 | 0 | Weather event |
| Total: | | 150 | 45 | 15 | 15 | 15 | |

As shown above, the analysis reviewed five windows, from the baseline schedule to month five, the as-built schedule. In the first window (month 1), the contractor is entitled to a 15-day excusable and compensable delay due to a differing site condition. In the third window (month 3), the contractor incurred a 15-day non-excusable subcontractor delay. In the fifth window (month 5), there is a 15-day excusable non-compensable delay due to a weather event. In sum, the project finishes 45 days late. Thus, the contractor is entitled to a 30-day time extension and 15 days of delay damages. The owner, on the other hand, is entitled to assess 15 days of liquidated damages against the contractor.

### C. Time Impact Analysis (TIA) (Prospective or Retrospective, Forward-Looking)

The time impact analysis (TIA) method of delay analysis is a widely accepted technique for measuring the impact that an excusable event, such as a scope modification or discovery of a differing site condition, has on the project completion date. Because it is a forward-looking approach, TIA is ideal for prospective analyses performed contemporaneous with the delay impact. In fact, many construction specifications require the use of the TIA method in prospective delay scenarios. The TIA is also viable as a forward-looking retrospective analysis and is widely accepted by courts and boards. Whether a time impact analysis is appropriate in a retrospective scenario depends on the availability of contemporaneous schedule data and the given circumstances of the impacts being studied.

The analysis is done by inserting the impact into the available project schedule in the form of a fragmentary network or "fragnet" and comparing the changes in the critical path. If enough float exists on the activities affected by the change, the project completion date will remain unchanged after insertion of the fragnet; if it does not, then the project will experience a delay and show a later completion date. The quantified delay attributable to the impact is the variance in completion date after insertion of the impact's fragnet.

The TIA approach is broken down into five general steps. First, the contractor identifies an excusable delay impact (i.e., owner design issue, etc.). Second, the contractor identifies the most recently updated project schedule prior to identification of the delay event, which represents the unimpacted schedule.[2] Third, the contractor creates the fragnet detailing the delay impact activities (i.e., procure additional materials, install additional work, etc.). When a TIA is performed prospectively, the fragnet duration is an estimate as the full extent of the delay event is not known. When performed retrospectively, the fragnet duration aligns with the actual as-built dates for the impact. Fourth, the contractor inserts the fragnet into a copy of the unimpacted schedule and reruns the CPM calculations to create the impacted schedule. Fifth, the contractor compares the forecasted completion dates between the unimpacted schedule and the impacted schedule. Any variance between the two schedules is the duration of the critical path impact attributable to the excusable delay event.

In performing a TIA, best practice states that the schedule chosen as the unimpacted schedule should be the most recent update prior to the date of the impact event. Using the most recent update prior to the delay event as the unimpacted schedule ensures the analysis considers the status of the overall project at the time of the impact. Oftentimes, however, contemporaneous schedule updates are not available and only a baseline schedule exists to serve as the unimpacted schedule for the analysis. In such instances, the TIA is sometimes referred to as an "Impacted As-Planned" analysis. The mechanics of an Impacted As-Planned are identical to a TIA – the only difference being the chosen unimpacted schedule.

**Example – *Time Impact Analysis***
During month four of the construction of a building project, the owner elects to change the floor finishes from a readily available carpet to a specialty ceramic tile that has a long lead time. The contractor develops a fragnet that includes three activities: (1) purchase of the specialty tile from distributor (15 days); (2) procurement of the tile from Italy (30 days); and (3) installation of the specialty tile (15 days). Overall, the fragnet duration is 60 days.

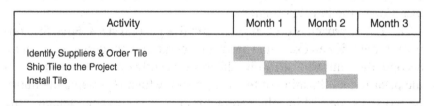

| Activity | Month 1 | Month 2 | Month 3 |
|---|---|---|---|
| Identify Suppliers & Order Tile | | | |
| Ship Tile to the Project | | | |
| Install Tile | | | |

Immediately prior to the initiation of the change by owner the project schedule showed substantial completion would be achieved at the end of month four. Because the owner directed this change in month four of the project, the contractor identifies its month three schedule update as the unimpacted schedule. The month three schedule notes the project completion date is scheduled to occur at the end of month four. The contractor then inserts the fragnet of activities into the month three schedule and updates it, in order to identify the anticipated impact to the project completion date. In the figure, red activities are critical, blue activities are non-critical, and green activities represent delays.

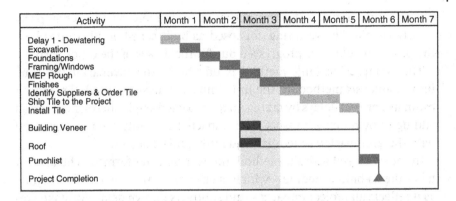

| Activity | Month 1 | Month 2 | Month 3 | Month 4 | Month 5 | Month 6 | Month 7 |
|---|---|---|---|---|---|---|---|
| Delay 1 - Dewatering | | | | | | | |
| Excavation | | | | | | | |
| Foundations | | | | | | | |
| Framing/Windows | | | | | | | |
| MEP Rough | | | | | | | |
| Finishes | | | | | | | |
| Identify Suppliers & Order Tile | | | | | | | |
| Ship Tile to the Project | | | | | | | |
| Install Tile | | | | | | | |
| Building Veneer | | | | | | | |
| Roof | | | | | | | |
| Punchlist | | | | | | | |
| Project Completion | | | | | | | |

As shown above, the TIA forecasts that the specialty tile change will push the project completion date from the end of month four to midway through month six, or a 45-day delay. Note that 15 days of the 60-day frag-net duration run concurrent with the critical finishes duration. Thus, the contractor is entitled to request an excusable compensable delay claim of 45 days.

### D. Collapsed As-Built Analysis (Retrospective, Backward-Looking)

The collapsed as-built delay analysis—also known as a but-for analysis—is a backward-looking retrospective technique that begins with the as-built schedule and then subtracts known delays to demonstrate the hypothetical project completion date but for the delays. The resulting "collapsed as-built" schedule demonstrates when a project theoretically would have been completed without the omitted delays.

The key principle of the collapsed as-built method is the demonstration of a project's completion without the effect of delays caused by another party. The philosophy is that without the other party's delays, the project would have been completed earlier and thus the claimant is entitled to a time extension for the difference between that but-for date and the actual completion.

A collapsed as-built schedule analysis starts by developing or refining the project's as-built schedule. Next, actual delay events caused by the claimant and respondent are identified. The delays attributable to one of the parties are removed from the as-built schedule, thereby "collapsing" the schedule and leaving the delays caused solely by the other party clearly visible. The resulting collapsed as-built schedule illustrates how the project would have progressed but-for the delays of the other party.

The collapsed as-built method is effectively the inverse of the time impact analysis method. In the time impact analysis method, the analyst estimates project delays by taking an unimpacted as-planned schedule and adding known delay impacts. The impacted schedule then is compared with the unimpacted schedule to quantify project delays.

In the collapsed as-built method, the reverse is performed. The analyst takes the as-built schedule—which is effectively an "impacted" schedule as it reflects all project impacts—and subtracts known delay events to create the "unimpacted" schedule. This unimpacted schedule is compared with the as-built schedule to quantify project delays.

Like the time impact analysis technique, the collapsed as-built method is easy to understand and present, and can provide consideration for concurrent delays if performed properly. However, it often requires considerable reworking of the as-built schedule to ensure all delays are considered and the underlying schedule data is valid, which can be a challenging, time-intensive process.

Courts are mixed on their acceptance of the collapsed as-built technique. Generally, this method is applied in cases where reliable as-built schedule information exists, but contemporaneous schedule updates either do not exist or are not reliable. Other schedule analysis models rely heavily on this data so when it is not available, the collapsed as-built approach provides a path forward. A lack of contemporaneous information, however, also lends itself to criticism of the as-built schedule developed by the claimant during the analysis. Therefore, it is important to maintain strict guidelines and thoroughly consider all available project documentation.

**Example – *Collapsed As-Built Analysis***
The contractor was under contract with the owner to construct a building project within 135 days, or 4.5 months. During the finishes portion of the work, the owner added an extensive millwork package to the contractor's scope of work. The contractor's as-built schedule is shown below. The activity in green is the additional millwork. The contractor was able to complete the work within 165 days, or 5.5 months.

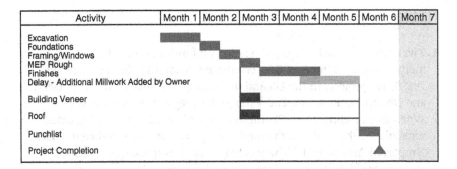

When the as-built schedule is collapsed by removing the millwork activity, the project completion date is pulled back to 135 days, or 4.5 months; therefore, the calculated delay attributable to the added millwork is 30 days, or 1 month.

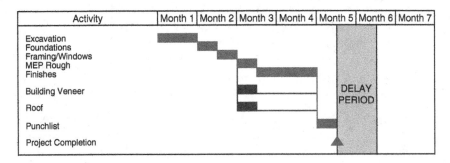

# VII. Summary

The four forensic scheduling methodologies reviewed above are commonly used by claimants to quantify schedule impacts and establish entitlement for excusable delays (both compensable and non-compensable). Respondents often use these same methodologies as well to rebut affirmative claims and to establish that asserted delay claims are non-excusable, non-compensable, or simply overstated. The choice of which method is appropriate depends on a number of factors such as contract provisions, the nature of the delays being analyzed, availability of reliable schedule information, and other considerations.

## Notes

**1** The most commonly cited publication on forensic scheduling methodologies is the AACE International Recommended Practice No. 29R-03 for projects in the US and the Society of Construction Law Delay and Disruption Protocol, 2nd Version, for projects outside the US.

**2** When the Baseline or As-Planned schedule is used as the unimpacted schedule for the analysis, the methodology is sometimes referred to as an "Impacted As-Planned." The mechanics of an Impacted As-Planned analysis are identical to a Time Impact Analysis.

# 7

# Step 6: Calculate Damages

After the claimant proves entitlement for its claim against the respondent and it determines if the issue caused an excusable delay to the claimant's work, the claimant should then move to calculate its damages resultant from the described issue(s). For claimants that are contractors or sub-contractors, there are five main categories of damages, which include: (1) Scope Change Damages; (2) Productivity Damages; (3) Acceleration Damages; (4) Delay Damages; and (5) Consequential Damages. Individual claims can include one or more of these damage categories. Note that it is important to review the subject contract between the claimant and the respondent because certain damage categories might be explicitly waived therein, such as consequential damages. For the purpose of this chapter, the claimant is a contractor, and the respondent is an owner.

The first three damage categories (Scope Change Damages, Productivity Damages, and Acceleration Damages) are typically made up of hard construction costs and allowable contractor markups. Hard construction costs are referred to as "brick-and-mortar costs" that relate to subcontractors, vendors, materials, equipment, and self-performed labor, which all relate to the physical incorporation of specific work at the project.

The fourth damage category, Delay Damages, includes time-related costs and allowable contractor markups. These time-related costs do not get incorporated into the work and are not related to any one on-site construction activity. Delay damages can include direct project overhead costs for extended general conditions, extended equipment costs, and general labor, such as maintenance of traffic, or cleanup. The fifth and final damage category, Consequential Damages, relates to costs not associated with on-site project activity, such as home office overhead, loss of financing, business and reputation, impacts, and lost profits.

Consequential damages are not subject to markups and are frequently mutually waived within contract agreements. Consequential damages that an owner might incur include loss of rental expenses, loss of use, loss of income and profit, extended financing, loss of business and reputation, and loss of management or employee productivity.

Depending on what cost information exists, damages for the first four categories may be calculated by one of four methods: (1) Actual Cost Method; (2) Agreed Upon Cost Method; (3) Estimated Cost Method; or (4) Modified Total Cost Method. The following is a review of each of these four methods followed by a detailed review and examples for each damage category.

## I.  Methods to Calculate Damages

### A.  Actual Cost Method

Actual damages are the direct costs that the contractor incurred as the result of an impact. This method is most preferred in retrospective claims, where the contractor has already sustained the resultant damages and is able to provide discrete expenses associated with an impact. Damages associated with productivity loss may be difficult to calculate using an actual cost method, as these inefficiency costs may be co-mingled with the "efficient" component of the work.

Actual cost damages are typically captured within the contractor's job cost report under dedicated cost codes or are otherwise segregated. In addition to a segregated listing of actual expenses, the contractor should ensure the costs are auditable and provide backup for each expense to the extent possible. For contractor labor charges, payroll records should be coupled with timesheets or daily reports, where possible. Subcontractor or vendor charges should be supported by invoices that detail the work performed and/or materials and equipment supplied.

Actual costs are often scrutinized and alleged to be excessive. Soliciting a subcontractor's proposal or presenting a vendor's actual costs does not mean these costs represent a reasonable cost for the work; thus, it is sometimes necessary to present alternate pricing that was obtained at the time or compare these costs with industry standard cost data to counter such an attack.

Equipment is a difficult cost component to access actual costs for. For rented equipment, the contractor can reference vendor invoices, which typically define a weekly or monthly rate. Depending on the duration of the impact, it might be necessary to calculate an hourly, daily, or weekly rate based on a monthly invoice charge. For instance, if the invoice for a rented scissor lift notes a monthly rate, and the contractor used the lift to install additional ceiling work for a one-week time period and used the equipment elsewhere on site during other times, the contractor would simply calculate a weekly rate for the equipment and charge this amount in its claim.

For contractor-owned equipment, rates are sometimes defined within the contract. Equipment rates can also be evaluated via equipment cost software such as EquipmentWatch. Construction cost software such as RSMeans also includes equipment rates for certain items. Lastly, rental rates for equipment can be obtained via equipment rental shops that post rental rates. When a contract stipulates equipment rates, and the equipment claim is significant, it is best to compare the rates noted within the agreement against fair market value pricing obtained by these other sources to ensure the stipulated rates are reasonable, because the parties to the contract typically assume the rates will be used on smaller change orders and these rates should not create a windfall for the claimant on larger claims.

## B. Agreed Upon Cost Method

When contracts include rates sheets for labor, equipment, and possibly other items, it is sometimes appropriate to utilize such rates in small claims. It is necessary to support the time related to each component through time sheets and/or daily reports. However, on larger claims, it is often best to use actual costs, if possible, in order to defend against a windfall rebuttal. For instance, if a rate of $60.00 per hour is established for a common laborer, and it turns out the actual cost for such laborer is $40.00 per hour, it can be argued that utilizing the higher rate is excessive and creates a windfall for the claimant, particularly if markup is applied on top of the higher rate, regardless of its inclusion in the contract. See Figure 7.1.

For unit price contracts or for contracts that stipulate unit prices for certain additional work, it is appropriate to utilize such rates. The critical step

2.    Labor directly supplied by the Contractor shall be charged, including Contractor's mark-up, unless otherwise specified in scope of work as follows:

    a)    Supervisors will be billed @$85 per hour

    b)    Carpenter's labor will be billed @$65 per hour

    c)    Carpenter's helpers will be billed @$45 per hour

    d)    Unskilled labor will be billed @$30 per hour

    e)    Electrical labor will be billed @$80 per hour

**Figure 7.1**    Excerpt of contract executed between framing contractor and owner for renovation of existing residence in Denver, CO.

here is to define the added quantities via surveys or detailed takeoffs. Certain unit price contracts include a provision whereby the contractor can renegotiate the unit price rate for a work item if a quantity overrun results from the work, such as 125% over the estimated quantity figure. This can be important if the contractor has an unusually low unit price for a claim item and the overrun work would result in a loss to the contractor based on the stipulated rate. See Figure 7.2.

## C.    Estimated Cost Method

The estimated cost method is widely used for calculating damages for prospective claims or if the contractor did not or could not segregate its actual costs related to the claim. There are several methods for estimating impact costs.

### 1.    Subcontractor/Vendor Estimates

Contractors often rely on subcontractor and vendor proposals to price additional work. Fair market pricing may not be properly established by simply soliciting a single source bid for the subject work so it is good practice for contractors to verify subcontractor and vendor pricing through quantity takeoffs and a review of subcontractor/vendor unit prices against industry standard cost data, and any other data that may be applicable. The process of negotiating changes between an owner and a contractor can be time-intensive, so when the contractor takes the extra step to verify and substantiate subcontractor and vendor pricing, it generally goes a long way with the owner.

| BID ITEM #VOYA SECTION | Mn/DOT SECTION | ITEM | QUAN-TITY | UNIT OF MEASURE | UNIT PRICE | AMOUNT OF BID |
|---|---|---|---|---|---|---|
| 1 VOYA 01025 | Mn/DOT 2021 | Mobilization | 1.00 | Lump Sum @ | $33,000.00 = | $33,000.00 |
| 2 VOYA 01025 | Mn/DOT 2575 | Temporary Access Drives | 10.00 | Each @ | $1,800.00 = | $18,000.00 |
| 3 VOYA 01560 | Mn/DOT 1710 | Barriers and Traffic Controls | 1.00 | Lump Sum @ | $9,000.00 = | $9,000.00 |
| 4 VOYA 01050 | Mn/DOT 1508 | Construction Surveying and Staking | 1.00 | Lump Sum @ | $30,000.00 = | $30,000.00 |
| 5 VOYA 02110 | MnDOT 2101.501 | Clearing | 7.20 | Acre @ | $3,472.00 = | $24,998.40 |
| 6 VOYA 02110 | MnDOT 2101.507 | Individual Stump Removal | 50.00 | Each @ | $75.00 = | $3,750.00 |
| | MnDOT 2573 | Storm Water Erosion | | | | |
| VOYA 02215 | MnDOT 2575 | and Sediment Controls | | | | |
| 7 | MnDOT 2573.540 | Filter Log Bioroll Straw 7" | 1,770.00 | Ln. Ft. @ | $4.70 = | $8,319.00 |
| 8 | MnDOT 2573.540 | Filter Log Type Compost 10" | 1,482.00 | Ln. Ft. @ | $6.60 = | $9,781.20 |
| 9 | MnDOT 2575.525 | Erosion Blanket Cat. 3 | 5,183.00 | Sq. Yd. @ | $1.85 = | $9,588.55 |
| VOYA 02372 | MnDOT 2511 | Rip Rap | | | | |
| 10 | MnDOT 2511.502 | Imported Rip Rap Class III | 230.00 | Tons @ | $15.00 = | $3,450.00 |
| 11 | MnDOT 2511.504 | Quarry Run Rip Rap Class I | 115.00 | Tons @ | $16.00 = | $1,840.00 |

**Figure 7.2** Excerpt of unit price contract executed between contractor and owner for construction of a new paved bike trail.

## 2. Cost Estimating

Cost estimating involves the takeoff of quantities and the identification of reasonable unit prices related to the subject work. The estimated quantities are then multiplied by the unit prices to estimate the direct costs related to the work. Once direct costs are calculated, indirect costs and markups are added to the direct costs to arrive at a claim amount.

### a. Quantity Takeoffs

In order to estimate an additional or changed scope of work, it is necessary to determine the quantities of additional or changed work. The use of digital takeoff software is recommended because it can provide the owner with a graphical representation of the locations of changed work, as well as the linear footage, area, or volume of claimed work directly on the subject plans. Popular digital takeoff software includes OnCenter's On-Screen Takeoff (Figure 7.3), Bluebeam's Revu, and Autodesk's Revit. Hand takeoffs are also used but may make verification of an estimator's work more cumbersome.

### b. Unit Cost Pricing for Direct Costs

**i. Industry Standard Construction Cost Data** An important aspect of estimating claim-related work is to ensure cost backup has sufficient basis. The use of industry standard construction cost data that is localized to the project area is a widely accepted method to establish fair-market unit pricing. Widely used construction cost software includes, but is not limited to, RSMeans, Xactimate, and EquipmentWatch. In addition to general estimating databases, industry-specific estimating software is available for certain trades, such as Trimble's TRA-SER, which includes pricing databases for MEP trades.

RSMeans is an industry leader in the construction cost industry. Engineering and construction students are often trained to use RSMeans in undergraduate programs. RSMeans localizes unit prices by Country (the US or Canada), State/Region, and then City. RSMeans is applicable for both new construction as well as renovation/reconstruction work. Xactimate is an industry leader in the restoration cost data marketplace. It is widely used by insurance claims adjusters and restoration contractors to estimate work. Accordingly, it is very effective for restoration/reconstruction work that involves removals and replacements. EquipmentWatch is a leading source for equipment costs. It provides hourly, daily, weekly, and monthly rates for thousands of pieces of equipment. In addition, EquipmentWatch provides equipment use rates and standby rates, which can be important for cost recovery.

When utilizing industry standard cost data, it is important to evaluate projects to determine if adjustments are necessary to the data. For instance, if the quantity count is large, there is a good chance that the fair market unit price for the work will be lower than what is noted as an

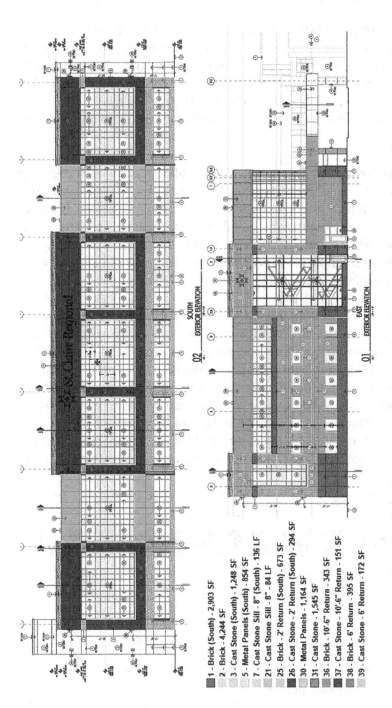

1 - Brick (South) - 2,903 SF
2 - Brick - 4,244 SF
3 - Cast Stone (South) - 1,248 SF
5 - Metal Panels (South) - 854 SF
7 - Cast Stone Sill - 8" (South) - 136 LF
21 - Cast Stone Sill - 8" - 84 LF
25 - Brick - 2' Return (South) - 673 SF
26 - Cast Stone - 2' Return (South) - 294 SF
30 - Metal Panels - 1,164 SF
31 - Cast Stone - 1,545 SF
36 - Brick - 10'-6" Return - 343 SF
37 - Cast Stone - 10'-6" Return - 151 SF
38 - Brick - 6' Return - 395 SF
39 - Cast Stone - 6' Return - 172 SF

**Figure 7.3** Excerpt of On-Screen Takeoff showing the total quantity of concrete and formwork placed by contractor at the foundation level for use in the estimation of the value of work completed prior to contract termination by the owner for cause.

industry standard due to economies of scale. Similarly, if a quantity count is small, an industry standard unit price might need upward adjustment to account for minimum charges that apply to most work. Moreover, special circumstances, such as project security requirements and accessibility, may require the unit prices to be adjusted. In sum, any type of adjustment to an industry standard unit price should be clarified within the cost estimate.

**ii.  Public Agency Unit Price Data**  Many public agencies, such as certain state departments of transportation, publish pricing data obtained from contractor proposals on various projects. If the subject project is similar in nature to projects with listed cost data, these can be a useful estimating and cross-referencing source. When using this data, it is best to average the unit costs for multiple projects and multiple contractors. Figures 7.4 and 7.5 are examples of such pricing data.

**iii.  Contractor Historical Cost Data**  Contractor historical cost data can also be used to establish the reasonableness of cost inputs of claims; however, finders of fact will often give more weight to industry standard cost data, as it is an independent source, rather than in-house proprietary figures that are not relied upon by others. If such data is used, the contractor should establish that the historical cost data was generated from similar projects. For instance, if the subject project is a multi-family development, the similar project would ideally be a multi-family project. This type of analysis is common when industry standard cost data is not

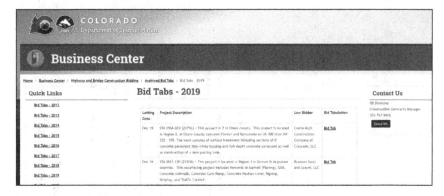

**Figure 7.4**  Pricing data on bid tabs.

| Item Code | Description | Quantity | | (0) -EST-<br>Engineer's Estimate | | | | | |
|---|---|---|---|---|---|---|---|---|---|
| | | | | Unit Price | Amount | Unit Price | Amount | Unit Price | Amount |
| SECTION: | 0001 BID ITEMS | | | | | | LCC: | | |
| 201-00000 | Clearing and Grubbing | 1.000 | L S | 50,000.00000 | 50,000.00 | 5,000.01000 | 5,000.01 | 195,400.00000 | 195,400.00 |
| 202-00033 | Removal of Pipe | 20.000 | EACH | 1,500.00000 | 30,000.00 | 575.00000 | 11,500.00 | 751.00000 | 15,020.00 |
| 202-00090 | Removal of Delineator | 139.000 | EACH | 7.00000 | 973.00 | 18.99000 | 2,639.61 | 28.00000 | 3,892.00 |
| 202-00220 | Removal of Asphalt Mat | 30,201.000 | SY | 6.50000 | 196,306.50 | 4.55000 | 137,414.55 | 6.00000 | 181,206.00 |
| 202-00240 | Removal of Asphalt Mat (Planing) | 143,575.000 | SY | 2.50000 | 358,937.50 | 1.28000 | 183,776.00 | 4.00000 | 574,300.00 |
| 202-00246 | Removal of Asphalt Mat (Planing) (Special) | 411.000 | SY | 15.00000 | 6,165.00 | 19.49000 | 8,010.39 | 4.00000 | 1,644.00 |
| 202-00250 | Removal of Pavement Marking | 500.000 | SF | 0.50000 | 250.00 | 7.50000 | 3,750.00 | 11.00000 | 5,500.00 |
| 202-00453 | Removal of Portions of Present Structure (Class 2) | 82.000 | SY | 250.00000 | 20,500.00 | 330.00000 | 27,060.00 | 2,040.00000 | 167,280.00 |

Colorado Department Of Transportation — Printed On: 12/23/2019 — Tabulation of Bids — Page 1 of 14 — Contract ID:

**Figure 7.5** Project costs.

available for certain types of specialty work. If industry standard data is available, it is prudent for the claimant to cross-reference historical cost data against industry standard pricing to establish the reasonableness of the claimed cost items.

### 3. Industry Studies and Scholarly Research Papers

Industry studies may consist of specialty industry studies or general industry studies. Specialty industry studies are mostly commissioned by construction associations and organizations and are typically based on data compiled from actual construction projects. Some such studies measure the effects of acceleration, learning curve, overtime, and weather effects, among others. Most of these subject-specific productivity studies are either peer-reviewed scientific articles written on factors affecting labor productivity in construction projects or studies published by recognized labor associations and industry groups (Business Roundtable, Construction Industry Institute, etc.).

General industry studies are typically used when specialized studies are not applicable and when sufficient contemporaneous and project-specific documentation (such as detailed and/or reliable labor and production tracking records) do not exist to demonstrate the productivity loss. Calculations relying on general industry studies are subject to additional scrutiny because they are not project- or subject-specific and thus are less demonstrably applicable to the situation giving rise to the claim being prepared. An example of a general industry study is the productivity loss factors established by the Mechanical Contractors Association of America, Inc. (MCAA). MCAA provides a list of 16 factors which affect productivity, and identifies percentages for severity, which are either minor, average, or severe. Application of these loss factors in accordance with the MCAA's guidance generates estimated labor productivity losses. These studies are often criticized as they provide a very simplified "one-size-fits-all" approach to loss productivity estimates and are calculated based on the analyst's discretionary use of the applicable factors and severity of those factors. Accordingly, this type of analysis may best be used to support another cost calculation technique.

Academic research papers have also addressed the calculation of lost productivity damages. Dr. William Ibbs, a professor at the University of California at Berkley, has produced extensive writings on this subject that reference productivity graphs for crowding, extended overtime, and shift work. Like industry association reports, this is a one-size-fits-all position that is often criticized. While industry studies and scholarly research papers may be used for standalone calculations, they often supplement and support other cost methodologies, rather than for the exclusive use to estimate claim damages.

### 4. Measured Mile Analysis

A measured mile method is a favored way to calculate productivity damages, as it utilizes project-specific information and considers actual contract performance rather than relying on estimates. It measures the labor productivity of the contractor on an unimpacted area of work against the productivity on a similar impacted area of work. The difference between the two is the loss of labor productivity. While this technique is favored, it is difficult to apply as it requires detailed reporting and an unimpacted portion of the work to serve as the baseline. Having these elements available on a building project is not always possible. It is more common to use this technique on linear projects, such as roadway or utility improvements. When it is difficult to locate an unimpacted area of work because the impact affected all areas, the claimant can evaluate the productivity of a least impacted area of work and utilize this within a measured mile analysis.

For example, a utilities contractor (claimant) that has a contract with the owner (respondent) to install 5,000 feet of 8" waterline has a productivity

**Table 7.1**  Estimate of contractor's productivity

| Location | Output (Quantity) | Input (Hours) | Productivity Rate | Status | "Should Have Spent" Hours |
|---|---|---|---|---|---|
| Sta 00+00 to 20+10 | 2,010 linear feet | 550 hours | 3.65 LF/hour | Unimpacted | |
| Sta 20+10 to 30+10 | 1,000 linear feet | 465 hours | 2.15 LF/hour | Impacted | 270 hours |
| Sta 30+10 to 50+00 | 1,990 linear feet | 530 hours | 3.75 LF/hr | Unimpacted | |

claim due to unforeseen site conditions between Station 20+10 and Station 30+10 (representing 1,000 feet of the waterline). The claimant proved its damages through a measured mile analysis by performing productivity calculations for unimpacted sections of the waterline and comparing them to the productivity for the area that was impacted. The contractor's productivity was calculated as shown in Table 7.1.

The claimant's damages are calculated by taking the differential in labor expended in the impacted area, less than the hours which should be incurred based on the contractor's actual productivity in unimpacted areas of the project. The contractor's average productivity in the unimpacted areas is calculated by taking the unimpacted footage (2,010 linear feet + 1,990 linear feet = 4,000 linear feet) and dividing this by the unimpacted labor hours (550 hours + 530 hours = 1,080 hours), which yields 3.70 linear feet per hour. Thus, the contractor should have spent 270 hours in the impacted area (1,000 linear feet / 3.7 = 270 hours), but it spent 465 hours, so the productivity loss was 195 hours (465 hours – 270 hours). The calculated hours lost is then multiplied by the average labor rate to calculate the cost of the productivity loss.

### 5.  Earned Value Analysis

The earned value method is another project-specific methodology to estimate productivity damages. This method often relies on payment applications between the claimant and respondent, the claimant's estimated hours for the work, and the claimant's job cost reports that identify the actual number of labor hours expended over periods of time. For instance, if a mechanical subcontractor (claimant) that has a lump sum subcontract with a general contractor (respondent) on a two-story office building project asserts loss of productivity damages against the general contractor for impacts to the rough mechanical work on the second floor of the building, it could potentially create an earned value analysis to estimate its productivity loss. For this hypothetical claim, assume that the mechanical subcontractor started the rough mechanical work on the first and second floors at the same time and the floor plates are very similar, and that the impacts took place on the second floor for the first four months of rough mechanical work.

To compare the productivity rate the claimant achieved on the unimpacted rough mechanical work on the first floor for the first four months against the productivity on the impacted second floor work, the claimant would first calculate its earned hours. To calculate earned hours through month four, the claimant would multiply the percentage complete of the rough mechanical work for each floor listed in the subcontractor-general contractor payment application against the claimant's total budget hours for the rough mechanical work for each floor. If the claimant estimated a total of 5,000 hours for each floor for rough mechanical work, and the month four payment application noted 50 percent complete on the first floor and 20 percent complete on the second floor, the earned hours would be 2,500 (5,000 × 50%) on the first floor and 1,000 (5,000 × 20%) on the second floor. Then, the claimant would refer to its job cost report to identify the actual hours that it incurred on the first and second floors, and compare them to the earned hours. If the job cost report noted 2,580 hours on the first floor and 2,108 hours on the second floor, the productivity rate would be 96.9 percent on the first floor (2,500 hours / 2,580 hours) and 47.4 percent on the second floor (1,000 / 2,108). The actual loss would be the difference in hours between the anticipated hours of 1,032 (1,000 hours / 96.9 percent) for the second floor against the actual hours incurred of 2,108, or 1,076 hours. The 1,076 hours would then be multiplied times the average actual labor rate to estimate the damages.

There are several considerations, for both the earned hours and actual hours, which the claimant should check to present a robust earned value analysis. The claimant should confirm the labor estimate is reasonable and properly allocated, as they are part of the basis of the earned value. This would mean ensuring that the payment application did not front-load early line items more than later items with similar labor requirements, and that the allocations properly accounted for areas with unique or particularly challenging work that would require additional effort. How the percentage complete values were established should also be considered. With respect to actual hours, the claimant should confirm that it properly accounted for change order hours, and that the actual hours expended were properly recorded to the appropriate activity or line item.

### 6. Comparable Project Methodology

If the claimant is unable to utilize a measured-mile analysis or earned value analysis to calculate productivity damages, a comparable project methodology is sometimes used. A comparable project study compares the claimant's estimate of productivity on the impacted portion of work and compares it to a non-impacted portion of work on a similar project that the claimant has completed. Generally, the similar project is of similar size, is in a similar location, and is a similar project type. This effectively substitutes the standard measured mile for one based on a similar but not identical scope of work on a different project. This analysis may be met with skepticism given the variables and factors that inevitably differ between the comparable project and the subject project. As such, comparable project studies may be a good methodology to supplement another calculation method.

## D. Modified Total Cost Method

The modified total cost method is best described by first explaining the total cost method. The total cost method is often used when the contractor is unable to precisely calculate the damages related to an impact or impacts. The total cost method simply subtracts at-bid estimated costs from the total actual cost the contractor incurred on the project. Unlike the total cost method, the modified total cost method then reduces this amount by costs associated with any contractor bid error and performance issues.

The total cost method does not require the claimant to distinguish impact costs from unimpacted contract costs and it assumes all costs were reasonable and the contractor was not responsible for increased costs. Courts consider the total cost method to be a method of last resort and limit consideration to when the claimant can pass a four-part element test, meaning if the claimant fails one part of the test, it fails the entire test.

Four Part Total Cost Test:

1) **Contractor must prove it had no alternative method of calculating damages based on the circumstances.** If the contractor was reasonably able to track impact costs, but simply chose not to, it fails this test. Also, when the contractor has the information at its disposal to easily isolate the impact costs, but fails to do so and elects to assert a total cost claim, it fails this test.

2) **Contractor's original bid (at-bid estimate) was reasonable.** This can be done by comparing the contractor's bid against other bids for the subject project. If no or limited bids are available for comparison, the contractor can use an independent estimate, industry standard cost data and/or the contractor's historical pricing on similar projects.

3) **The actual costs that contractor incurred were reasonable.** The contractor must make some showing that it performed the project in an efficient manner despite the actions of the owner. This element also requires the contractor to demonstrate a good faith effort to mitigate damages.

4) **Contractor was not responsible for the extra costs incurred.** If the owner can demonstrate that the contractor worked inefficiently or improperly performed certain portions of the work, this method fails. Note that making an adjustment for this element switches the total cost method to a modified total cost method.

Because it is nearly impossible for a contractor to perform flawlessly on a project, the total cost method is not recommended. A more reasonable approach is the modified total cost method. Here, the contractor adjusts the damages to account for any bid error (Element 2), unreasonable costs that the contractor incurred (Element 3), and costs associated with contractor-caused impacts (Element 4). Thus, this method attempts to adjust for the unrealistic output of the total cost method.

Use of the modified total cost method still requires fulfillment of the four-part test. Courts require this test on both total cost claims and modified total cost claims. In addition, the modified total cost method is not limited to overall project costs. If the contractor performed multiple scopes, but the impact issue relates only to painting work, the modified total cost method can be used on this distinct division of work. Here, the contractor would start the process by subtracting its estimated at-bid painting costs from the total painting costs.

Considerable case law exists regarding total cost claims and modified total cost claims, so contractors should consult with qualified construction law attorneys to understand the applicable state or federal case law holdings related to modified total cost claims.

## II. Markup on Damages

Contracts often stipulate allowable markups on direct construction costs related to additional work. If no markups are stipulated therein, the contractor can refer to industry standard contractor markup for the contractor's fee, bonding, and insurance. Typical contractor fee markup fluctuates with the size of a project—the larger the project, the lower the markup percentage. As shown in Table 7.2, RSMeans notes that typical construction management fee percentages get lower as the volume of the project gets larger. For standard builders' risk and liability insurance, the average markup is typically around 1–2% of direct costs. For surety bonds, the markup generally runs between 0.5 and 2% of direct costs.

**Example –** *Markups*
The contractor's direct costs relating to an impact issue is $1.75M. The contractor-owner agreement notes that the contractor is entitled to the following markups for additional work:

- fee markup of 3.5% of the direct cost of the work;
- insurance markup of 1.1% of the direct cost of the work; and
- surety bond markup of 0.06% of the overall cost of additional work.

**Table 7.2**  Project size and fee markup (%)

| Project size ($) | Fee markup (%) |
| --- | --- |
| • 0–100,000 | 10 |
| • 100,000–250,000 | 9 |
| • 250,000–1M | 6 |
| • 1M–5M | 5 |
| • 5M–10M | 4 |
| • 10M–50M | 4 |
| • 50M+ | 2.5–4 |

Hence, the contractor calculates its markup costs as follows:

| | |
|---|---|
| Direct Costs: | $1.750,000 |
| Fee Markup: | $1.750,000 x 0.035 = $61,250 |
| Insurance Markup: | $1.750.000 x 0.011 = $19,250 |
| Direct Costs + Fee & Insurance Markup: | $1,830,500 |
| Surety Bond Markup: | $1,830,500 x 0.006 = $10,983 |
| Total Claim Cost: | **$1,841,483** |

## III. Damage Categories

As noted above, the five types of disputes that contractors typically make against owners can cause five types of damages:

1) Scope Change Damages
2) Delay Damages
3) Productivity Damages
4) Acceleration Damages
5) Consequential Damages

It is common for contractors to incur one or more of these damage types on a discrete impact. For instance, if the contractor has an agreement to renovate a historic building and the contractor identifies undisclosed asbestos-containing material (ACM) in vinyl floor mastic that covers a large part of the upper floors of the building, then this type of claim could result in multiple types of damages.

In this ACM example, damages might include scope change damages such as retaining an environmental abatement company to remove and properly dispose of the subject floor tiles. Delay damages might include extended general conditions because this additional work impacted the critical path and extended the project duration. Productivity damages might include finish subcontractor claims resultant from unanticipated crowding of the upper floors. And acceleration damages might include the cost of overtime premium paid to certain finish trades to mitigate the duration of the impact.

In another example, the contractor damages might only include one type of damage. If the contractor has a contract with the owner to construct a water treatment plant and the contractor has a dispute with the owner on a small quantity of non-critical concrete flatwork, the damages might be limited to scope change damages. Here, the three other damage categories would not apply. Thus, it is important to evaluate each impact to determine which damage components are applicable.

## A.  Scope Change Damages

Scope change damages can result from various types of owner impacts. Contractors typically calculate scope change damages in four steps. First, the contractor defines the items of additional work and/or changed work. Second, the contractor estimates the quantity for each item of work. Third, the contractor assigns a cost to the additional quantities of work. And, fourth, the contractor offsets the costs by reasonable credits if certain work is no longer required due to the change. For instance, if a dispute involves the installation of an additional 100 linear feet of interior wall on a commercial office project, the listing of additional work items might include:

- Layout
- Light Gauge Steel Framing
- Mechanical, Electrical, and Plumbing Rough-In
- Drywall
- Interior Doors, Frames, and Trim
- Painting
- Mechanical, Electrical, and Plumbing Finishes

Next, the contractor would provide a quantity takeoff of the additional work, which yields:

- Layout: 4 hours for a layout team of two
- Light Gauge Steel Framing: 100 linear feet of wall assembly with studs at 16" on center
- Mechanical, Electrical, and Plumbing Rough-In: wiring to 20 wall outlets and 10 light switches; ductwork to penetrate the wall at 10 locations
- Drywall: 1,600 square feet of 5/8" drywall
- Interior Door, Frames, and Trim: 10 frames, 10 doors, 200 linear feet of baseboard
- Painting: 1,600 square feet of painting, primer plus two coats of paint
- Mechanical, Electrical, and Plumbing Finishes: 20 wall outlets, 10 light switches

Third, the cost of each line item of work for the additional scope must be calculated by using actual costs, agreed upon unit prices, or estimated costs. If the contractor is relying upon subcontractor or vendor pricing, it should be evaluated to confirm the amounts are reasonable. The more detailed the contractor can be, the better. If possible, line items should be broken down into labor, materials, equipment, and installer markup.

Fourth, the dispute here is over added work, not changed work, so there would be no deduction in this instance for work that is no longer required.

## B.  Delay Damages

Contractor delay damages are typically time-related general conditions costs that include, but are not limited to, the following:

- **Project management and coordination staff wages, per diem, and vehicles** (typical staff positions include, but are not limited to, project manager, superintendent, project engineer, scheduler, clerk, quality control manager, safety manager, general labor that performs safety and cleanup, etc.)
- **Equipment costs**
- **General and maintenance labor (cleanup; temporary road maintenance; erosion-and-sediment control maintenance; site security)**
- **Temporary utilities**
- **Construction facilities** (temporary office; storage; portable toilets)
- **Construction aids** (temporary hoists; cranes; scaffolding and platform; stairways; swing staging)
- **Temporary barricades** (tarps; winter protection; fences; walkways; signage)
- **Insurance and permits.**

Obtaining actual general conditions costs should not be difficult as this information should be segregated in the contractor's job cost report. For example, if an owner design issue extends a project by 30 days, direct delay damages would include extended general conditions.

If actual costs related to extended general conditions are not readily accessible, it is not uncommon for contractors to estimate general conditions through a schedule of values analysis. Here, the contractor would identify its general conditions budget line item within its approved schedule of values with the owner that is used for payment application requests.

The contractor would take this amount and divide it by the original contract duration to generate a cost per day, cost per week, or cost per month. This general conditions rate is then multiplied by the delay duration.

Another way to estimate extended general conditions is to use industry standard costs data from sources like RSMeans. This simply requires a listing of relevant general conditions line items and associated fair market unit costs for each item. The unit costs are then multiplied by the delay duration.

## C. Productivity Damages

When an owner's actions cause the contractor to suffer lower than reasonably anticipated productivity levels and this results in financial damage to the contractor, the contractor can assert productivity damages against the owner. Productivity relates to the amount of input, such as labor hours, it takes to perform a unit of output. For instance, assume a flooring contractor reasonably estimated that it would install 100 square foot of carpet per manhour and this is the productivity rate that it achieved on the first half of the project; however, due to the owner assuming joint occupancy and beginning installation of its furniture and equipment during the second half of the project, the contractor only placed, on average, 50 square feet of carpet per hour. The contractor's productivity claim would cover the added labor and equipment hours it incurred due to the access issues and challenges associated with working in proximity with the owner's forces.

The key element of a productivity claim is the establishment of a cause and effect relationship of the owner impact. Because it is difficult to forecast productivity-related damages, contractors typically issue productivity claims retrospectively and support entitlement opinions with the use of contemporaneous information, such as daily reports, letters, inspection records, photographs, time-lapse video, etc. Accordingly, it is important for the contractor to notice the owner of productivity impacts and note that the contractor will report the quantum of the impact once the impact has ceased.

Contractors predominantly prove productivity damages through estimated cost method techniques, including the measured mile analysis, earned value analysis, comparable project study, industry study factors, and modified total cost claims. The measured mile technique is the preferred method for calculating productivity damages, followed by the earned value analysis.

Per the Association for the Advancement in Cost Engineering (AACE), common productivity impacts include:

1) Absenteeism and the Missing Man Syndrome
2) Acceleration (directed or constructive)
3) Adverse or Unusually Severe Weather
4) Availability of Skilled Labor
5) Changes, Ripple Impact, Cumulative Impact of Multiple Changes and Rework
6) Competition for Craft Labor
7) Craft Turnover
8) Crowding of Labor or Stacking of Trades
9) Defective Engineering, Engineering Recycle and/or Rework
10) Dilution of Supervision
11) Excessive Overtime
12) Failure to Coordinate Trade Contractors, Subcontractors, and/or Vendors
13) Fatigue
14) Labor Relations and Labor Management Factors
15) Learning Curve
16) Material, Tools, and Equipment Shortages
17) Overmanning
18) Poor Morale of Craft Labor
19) Project Management Factors
20) Out-of-Sequence Work
21) Rework and Errors
22) Schedule Compression Impacts on Productivity
23) Site or Work Area Access Restrictions
24) Site Conditions
25) Untimely Approvals or Responses

Per the Mechanical Contractors Association of America (MCAA), typical productivity impacts include:

1) Stacking of Trades
2) Morale and Attitude
3) Reassignment of Manpower
4) Crew Size Inefficiency

5) Concurrent Operations
6) Dilution of Supervision
7) Learning Curve
8) Errors and Omissions
9) Beneficial Occupancy
10) Joint Occupancy
11) Site Access
12) Logistics
13) Fatigue
14) Ripple
15) Overtime
16) Season and Weather Change

Common productivity impacts reported by productivity expert, Dr. William Ibbs, include:

1) Project and Contract Factors
2) Location and Environment Factors
3) Project Team
4) Managerial Actions and Decisions During Project Execution
5) Disruptive Events and Signs on Sites
6) Human (Worker) Reactions
7) External Factors

## D.  Acceleration Damages

Acceleration claims generally include direct costs to expedite work. Such costs commonly include overtime premium, expedited shipment costs, and additional supervision associated with extended working hours. Acceleration can be voluntary, constructive, or directed. Voluntary acceleration takes place when the contractor elects to accelerate the work without making a claim prior to incurring the acceleration costs. Recovery for this type of acceleration is difficult due to lack of owner notice of a claim. Constructive acceleration occurs when the owner rejects the contractor's time extension request, which forces the contractor to accelerate the work to compress the schedule. This is the most common type of acceleration. Directed acceleration occurs when the owner directs the contractor to complete the work in advance of a planned or specific completion date.

Acceleration claims for overtime premium are almost always proven through the actual cost method. Costs for added supervision are generally addressed by agreed upon hourly rates, industry standard supervision rates, or with actual costs. Because productivity often decreases with extended overtime usage, a productivity claim may accompany an acceleration claim.

**Example – *Acceleration Claim***
The owner rejects the contractor's claim that an owner impact entitles the contractor to a 60-day time extension. Because the project completion date remains unchanged, the contractor is forced to accelerate the finish trades on the project to achieve an on-time completion. The contractor does this by authorizing overtime for three of its subcontractors. Collectively, the subcontractors incur $805k in overtime premium costs. The contractor collects timesheets and payroll reports from each of the subcontractors that provide actual costs for overtime. In addition, the contractor included its added $30k supervision costs it incurred during the overtime hours. The contractor used its agreed upon hourly rates for the additional supervisory costs it incurred to monitor the overtime effort.

| | |
|---|---|
| Painting Subcontractor | $290k in overtime premium, supported by timesheets and payroll records |
| Carpet Subcontractor | $275k in overtime premium, supported by timesheets and payroll records |
| Ceiling Subcontractor | $240k in overtime premium, supported by timesheets and payroll records |
| Contractor | $30k in added supervisory time based on a contractually stipulated rate of $75 per hour. |
| Subtotal of Acceleration Costs | **$835k** |

## E.  Consequential Damages

Consequential damages, unlike direct damages, such as Scope Change Damages, Delay Damages, Productivity Damages, and Acceleration Damages, do not result in an increase in the contractor's cost of work on the subject project. Rather, consequential damages are indirect costs related to an owner breach, such as unabsorbed home office overhead or lost opportunity damages for profits that could have been made on other projects, including due to diminished bonding capacity. Most standard contract forms include a mutual waiver of most consequential damages. Allowable consequential damages that contractors can claim under certain standard contractors include interest on late payments and lost profits in the event of certain termination scenarios. If no waiver exists in the subject contract agreement, the owner might be liable to the contractor for all damages—both direct and indirect.

**A201, §15.1.7 Waiver of Claims for Consequential Damages:** This standard AIA provision is a mutual waiver of consequential damages. Specifically, contractor waives home office labor and expenses, loss of financing, loss of business, and loss of reputation. However, this section preserves contractor's consequential damage claim for lost profits arising directly from the work, meaning in the event of a contractor's proper termination of the contract per Section 14.1, the contractor can claim lost profits on incomplete work. In addition, Section 13.5 allows contractor to claim interest on late payments per the terms of the A201.

**ConsensusDocs 200, Section 6.6 Limited Mutual Waiver of Consequential Damages:** This provision includes a mutual waiver of consequential damages. Specifically, the contractor waives any claims related to loss of business, loss of financing, loss of profits not related to the subject project, loss of bonding capacity, loss of reputation, or insolvency. Note that Section 11.5.3 allows the contractor to make a claim for lost profits if the contractor properly terminates the contract.

**EJCDC C-700, Section 13.05.C Costs Excluded:** This provision excludes home office administrative labor and expenses, and capital expenses. Per Section 16.03, the contractor is not entitled to lost profits on incomplete work or other economic loss under any circumstance. The contractor is entitled to interest late payments per Section 16.04.B.

## F. Home Office Overhead Claim

The most common consequential damage claimed by a contractor is for unabsorbed home office overhead (HOOH). HOOH differs from contractor's field office overhead (FOOH) as it covers company overhead costs that are associated with all company projects. Examples of HOOH include, but are not limited to, executive salaries, marketing, principal office rent, company insurance, accounting expenses, etc.

HOOH claims are typically estimated by the Eichleay Formula. The Eichleay Formula has been used since the 1960s to estimate "unabsorbed" HOOH related to an impact event. Contractors rely on project income to fund or "absorb" home office expenses. If a project is delayed, the contractor may not be able to perform work on other projects that would generate income to help absorb HOOH costs. The Eichleay Formula calculates HOOH costs as a function of a firm's revenue.

The first step is to divide the contractor's total billings on the subject project by the contractor's total company billings for the duration of the subject project period (original duration plus impact periods). This percentage is then multiplied by the contractor's overall HOOH costs for the project period, which yields the allocable HOOH to the project.

The second step is to divide the allocable HOOH by the contract performance period to determine a daily rate for HOOH. The third and final step is to multiply the daily HOOH rate times the delay days, which results in the unabsorbed HOOH amount.

$$\left( \frac{Total\ Contract\ Billings}{Total\ Company\ Billings\ over\ Contract\ Period} \right)$$

$$\times\ (HOOH\ for\ Contract\ Period) = Allocable\ HOOH\ to\ Project$$

$$\left( \frac{Allocable\ HOOH\ to\ Project}{Contract\ Performance\ Period} \right) = Daily\ HOOH\ for\ Contract$$

$$(Daily\ HOOH\ for\ Contract) \times (Delay\ Days) = Unabsorbed\ HOOH$$

Over the years, claimants have used several methods for calculating HOOH depending on contract requirements and court rulings.

**Example – *HOOH Claim***

The contractor enters a contract with an owner to complete a $30 million apartment project. The stipulated duration of work is 24 months. Due to owner impacts, the contract duration is extended for 12 months. The contractor's amended contract amount at the conclusion of the project is $38 million. The contract agreement between the parties does not include a mutual waiver of consequential damages. The Eichleay Formula is used to calculate the contractor's unabsorbed HOOH claim.

- Original Contract Value: $30M
- Amended Contract Value: $38M
- Original Duration: 24 months or 730 calendar days
- Revised Project Duration: 36 months or 1,095 calendar days
- Delay Period: 12 months or 365 calendar days
- Total Firm Revenue During the 36 Months: $150M
- Firm Home-Office Overhead During the 36 Months: $5.5M

| | |
|---|---|
| Step One: | ($38M / $150M) x $5.5M = $1.39M of HOOH allocated to the project |
| Step Two: | ($1.39M / 1,095 CDs) = $1,269 of HOOH per Day |
| Step Three: | $1,269 x 365 CDs = **$463,185 in unabsorbed HOOH** |

## G. Lost Profits Due to Loss of Bonding Capacity

Contractors that predominately focus on public sector work rely on their bonding capacity to procure work. The Miller Act of 1935 (40 U.S.C. Section 3131 to 3134) is a federal law that requires prime contractors on federal construction contracts greater than a certain amount to post payment and performance bonds that run in favor of the government. Currently, the Federal Acquisition Regulation Part 28.102-1(a) notes that payment and performance bonds are required on construction contracts with federal entities that exceed $150,000.[1] Similar state statutes that require payment and performance bonds for state-funded work are referred to as Little Miller Acts. All 50 states have passed some form of a Little Miller Act. That said, if an owner impact causes a reduction or temporary halt in the contractor's bonding program, it is foreseeable that this would cause a reduction in the contractor's profits to the extent of the reduction period.

Because this type of damage is often criticized for being speculative, it is important for a contractor-claimant to first establish that its ability to procure bonds was restricted. This can be done through correspondence from the contractor's broker or bonding company that outlines the restrictions or reductions in the contractor's bonding program. Second, the contractor should identify advertised projects that it would normally have bid for but could not because of its reduced bonding capacity. This can be done by highlighting advertised projects that the contractor would bid on, which are often listed on federal or state agency websites. Third, the contractor should establish that its average monthly revenue, and therefore average monthly profits, decreased as a result of the owner impact. This can be done with contractor income statements that go back several years.

### Example – *Lost Profits Due to Lack of Bonding*

The contractor works exclusively in the public sector; thus, the ability to procure payment and performance bonds is critical for survival. The contractor's annual revenue averages $150 million for the past three years and its profit margin averages 4%. As a result of a large owner impact on a particular state project, the contractor's bonding company cut off the contractor's bonding program. As a result, it took one month for the contractor to obtain a bonding line from a new surety provider, but the single and aggregate program offered by the new surety is only half of that of the previous bonding program. After six months, the new bonding company increased the contractor's bonding program to previous levels. During this seven-month period, the contractor's monthly revenue dropped by an average of 40%. Hence, the contractor's damages are calculated as follows:

| | |
|---|---|
| Contractor's Average Monthly Revenue: | $150M / 12 months = $12.5M |
| Contractor's Average Month Profit: | $12.5M * 4% profit = $500k |
| Loss of Profit Due to Bonding Capacity Issues: | $500k * 40% loss * 7 months = **$1.4M** |

## H.  Interest Claims

Standard contracts often allow interest on "late" payments. Late payments are typically defined as owner payments that are not made, through no fault of the contractor, within the payment time stipulated or required by statute.

In standard AIA contracts, Section 9.7 of the AIA A201 notes that the interest rate shall be the rate agreed upon in writing by the parties or, in the absence thereof, "at the legal rate prevailing from time to time at the place where the Project is located." The legal rate of interest varies from state to state. For instance, New York set its legal rate of interest at 9% (Civ. Prac. L. & R. §§5003, 5004). Delaware's legal rate of interest is 5% above the Federal Reserve rate, which makes it subject to fluctuations (Title 6, Commerce and Trade, Subtitle II, Chapter 23, §2301(a)).

To calculate the interest on an overdue invoice, the first step is to determine how many days the payment is late. If a payment is less than 31 days late, a simple interest calculation is proper. For payments that are more than a month late, a monthly compounding interest calculation is appropriate. Simple interest is calculated by multiplying the amount of the invoice times the interest rate divided by 360, and then multiplying that amount by the number of days late.[2]

**Example – *Simple Interest Calculation***
The contract stipulates that the owner pays the contractor within 30 days of the contractor's submission of $1,500,000 invoice but pays 45 days after submission. Thus, the contractor is entitled to 15 days of simple interest. The project is in New York State, so the legal rate of interest is 9%.

Formula:

$$I = Pr\ (d/360)$$

- I is amount of simple interest owing
- P is the amount of principal or invoice amount
- r is the applicable interest rate, expressed as a decimal (rate as % divided by 100)
- d is the number of days for which interest is being calculated 360 is the number of days in a year based on a 3-month and 90-day quarter. This is based on the federal government's prompt payment calculator formula.

Calculation: $\$1,500,000 * 0.09 * (15 / 360) = $ **$5,625 interest claim**

If a payment is more than a month late, it is proper to use a monthly compounding interest calculation, unless compounding is disallowed per contract. This formula is a bit more complex than the simple interest calculation.

Monthly Compounding Interest Formula:[3]
Formula:

$$I = P(1 + r/12)^n * (1 + (r * d/360)) - P$$

- I is the amount of compound interest owing
- P is the amount of principal or invoice amount
- r is the applicable interest rate expressed as a decimal (rate as % divided by 100)
- 12 is the number of months in a year
- n is the number of months
- d is the number of days for which interest is being calculated

360 is the number of days in a year based on a 3-month and 90-day quarter. This is based on the federal government's prompt payment calculator formula.

Note that the first half of the equation calculates compounded monthly interest. The second half of the equation calculates simple interest on any additional days beyond the number of months.

**Example – *Monthly Compounding Interest Calculation***
The contract stipulates that the owner pays the contractor within 30 days of the contractor's submission of $1,500,000 invoice but it pays 75 days after submission. Thus, the contractor is entitled to 1 month and 15 days of interest. The project is in New York State, so the legal rate of interest is 9%.

Formula:

$$P(1 + r/12)^n * (1 + (r/360 * d)) - P$$

P  = $1,500,000
r  = 9% or 0.09
n  = 1 month
d  = 15 days

Calculation:  $1,500,000(1 + 0.09/12) * (1 + (0.09/360 * 15)) − $1,500,000
= **$16,917.19**

## I.  Loss of Profit on Incomplete Work

Loss of profit on incomplete work on the project is covered in Chapter 11, Termination Claims.

## Notes

**1** See https://www.acquisition.gov/content/part-28-bonds-and-insurance# i1088557
**2** See https://www.fiscal.treasury.gov/prompt-payment/interest.html
**3** See https://www.fiscal.treasury.gov/prompt-payment/monthly-interest .html

# 8

## Step 7: Formatting and Packaging the Claim

## I.  Overall Claim Report Outline

Claims reports should flow from section to section. The following is a sample outline for a typical construction claim that is being prepared by a claimant, which is followed by a sample outline for a typical construction claim that is being prepared by a claimant's outside consultant/expert.

Outline for a Claim Prepared by the Claimant

Title Page
Table of Contents

    (I)  Executive Summary
   (II)  Entitlement Analysis
  (III)  Forensic Delay Analysis
  (IV)  Damage Calculations
   (V)  Conclusions

Attachments (Key Demonstratives, Key Exhibits, Etc.)

Outline for a Claim Prepared by the Claimant's Consultant/Expert

Title Page
Table of Contents

    (I)  Assignment
   (II)  Documents Reviewed
  (III)  Executive Summary
  (IV)  Background

  (V)  Entitlement Analysis
  (VI)  Forensic Delay Analysis
 (VII)  Damage Calculations
(VIII)  Conclusions

Attachments (CV, Rate Sheet, Key Demonstratives, Key Exhibits, Etc.)

## II.  Formatting

Because dispute resolution may ultimately lead to a binding court or arbitration decision, it makes sense to follow common formatting rules for legal pleadings. The Federal Rule of Appellate Procedure 32 contains detailed requirements for the production of briefs, motions, appendices, and other papers that will be presented to the judges. It makes sense for claimants to follow these rules when formatting claim reports, as lawyers are comfortable with this style. The following formatting recommendations generally follows these rules (Table 8.1).

**Table 8.1**  Formatting recommendations

| Feature | Definition |
| --- | --- |
| Font style | A serif or san-serif typeface should be used for the body of the report. Common serif typefaces are Times New Roman, Georgia, Palatino, and Garamond. San-serif typefaces include Arial, Calibri, Helvetica, and Tahoma. |
| Titles | Boldface may be used for emphasis |
| Font size | For the body of the report, font size should be at least 11-point type and no more than 12. Header font size should be the same size as the font in the body of the report. |
| Justification | Either fully justified text or left-justified text. Fully justified text creates clean, vertical margins on the left and right, while left-aligned text creates a clean, vertical left margin and a ragged right margin. On Microsoft Word, consider using the "Hyphenation" option when full justification is selected. |
| Paper size | 8.5 x 11 inch page size |
| Line spacing | Double-spaced, but quotations more than two lines long may be indented and single-spaced. Headings and footnotes may be single-spaced. |

*(Continued)*

**Table 8.1** (Continued)

| Feature | Definition |
| --- | --- |
| Margins | One inch on all four sides |
| Page numbers | Page numbers may be placed in the margin |
| Line numbers | In certain venues, line numbering is required |
| Signature | The author(s) should sign the cover of the report |
| Paragraph indenting | If the text is double-spaced, indent one-half inch for each new paragraph |
| Outline numbering | Per the MLA Handbook, the proper order of outline numbering is as follows: I. A. 1. a. i. (1) (a). See Table 8.2. |
| Bulleted lists | Use bulleted lists when the order of the items does not change the meaning |
| Numbered lists | Use numbered lists when the order of the items is important. For instance, the steps to building a house include foundation work, framing, etc. Here, the order of the items matters, so a numbered list is appropriate. |
| Embedded lists | Embedded lists include items within a sentence. For instance, "When you travel for work, you need your driver's license, suitcase, briefcase, and money." Do not go beyond four items when using embedded lists. In general, embedded lists are not appropriate when presenting technical information. Consider using numbers within embedded lists when appropriate. For example, "The plaintiff's expert's position lack merit for the following three reasons: (1) his methodology is flawed; (2) his source data is incorrect; and (3) there are mathematical errors in his calculations." |

## III.  Writing Style and Organization

### A.  Be Factual, Clear, and Unemotional

Construction claims should not follow mystery novel protocol, where the reader will not know the conclusion until the last page. The writer of a claim should provide an overall roadmap for the reader at the beginning of the claim and in every section of the report so the respondent knows the claimant's position up front and the report then provides an explanation and basis of how the claimant reached such conclusions. Moreover,

**Table 8.2** Hierarchy of order of outline numbering

| |
| --- |
| **I. First Level** |
|     **A. Second Level** |
|         **1. Third Level** |
|             **a. Fourth Level** |
|                 **i. Fifth Level** |
|                     **(1) Sixth Level** |
|                         **(a) Seventh Level** |

construction claims are often technical in nature and writers can lose readers with technical jargon that the average person does not understand. Accordingly, the writer should draft claims reports to a hypothetical eighth grade audience. If an average eighth grader can understand the report, the writer has done his or her job well.

In addition, the writer should write in an objective tone and be factual and emotionless. When too much emotion is written into the claim, the facts can sometimes be ignored. Moreover, respondents don't want to read emotional reports—the facts are much more powerful. Leave the emotions for the attorneys.

## B. Active Voice vs. Passive Voice

The claimant should write the claim narrative in the active voice, not the passive voice. The active voice is when the subject of a sentence performs the verb's action. Sentences in the active voice have a strong, direct, and clear tone. Examples of active voice include:

- The contractor placed the concrete slab.
- The owner suspended the masonry work.

The two examples above have basic active voice construction: subject, verb, and object. The subject contractor performs the action described by "placed." The subject owner performs the action described by "suspended." In other words, the subjects take action in their sentences.

To the contrary, a sentence is in the passive voice when the subject is acted on by the verb. For instance:

- The concrete slab was placed by the contractor.
- The masonry work was suspended by the owner.

Making the sentence passive flipped the structure and necessitated the preposition "by." Using the active voice conveys a strong tone and the passive voice is weaker.

## C. CRAC Method for Writing Organization

In terms of writing organization, the CRAC methodology works very well for construction claims. As discussed in Chapter 5, CRAC is an acronym used in the legal community that stands for: Conclusion, Rule, Analysis, and Conclusion. It functions as a methodology for presenting a legal analysis and it is particularly useful in presenting all narrative sections of a construction claim. This method starts with the conclusion, so it provides the claimant's position up front. Claimants should consider using this format for each section of their claim in order present the information in an organized, clear, and consistent manner.

The Conclusion section of an CRAC directly makes a request of the respondent. The Rule section is the statement of rules typically defined within the contract documents or industry standards, which the claimant must follow in order to recover from a claim. The Analysis section of the claim defines how the claimant followed the rules and the results of the analysis in a step-by-step fashion. The Conclusion summarizes the analysis and again makes a request of the respondent. The following are examples of the CRAC methodology for individual portions of a contract claim.

### Example – *CRAC Example: Type 1 Differing Site Condition Entitlement Narrative*
Acme Construction is entitled to recover time and cost impacts from the Owner because of the differing site condition that Acme encountered during the earthwork and foundation work phase of the Project. [CONCLUSION]. Per Section 6 of the Agreement, Acme is to provide the

Owner and the Architect notice of a differing site condition within 7 days of recognition of such a condition. Section 7 of the Agreement notes that a Type 1 Differing Site Condition occurs when the actual geotechnical conditions of the project materially differ from the reported geotechnical conditions within the Contract Documents, and when such condition occurs and proper notice is given, the contractor is entitled to recover reasonable cost and time impacts from the Owner, if any occur. [RULE]

Acme first observed the subject differing site condition on June 20th, shortly after it started excavation for the building foundation. Three days later, on June 23rd, Acme placed the Owner and the Architect on notice of this issue. A copy of this notice letter is attached herein. The Contract Drawings and Geotechnical Report, which are both Contract Documents under Section 2 of the Agreement, explicitly identify the subsurface condition of the Project to be silty sand. Below are excerpts of both documents that confirm this anticipated condition, upon which Acme based its bid. On June 23rd, Acme's excavation subcontractor encountered a rock outcropping just below the surface from Column Line A1 to Column Line A4. Below are photographs of this condition. [ANALYSIS]

The noted rock outcropping represents a Type 1 Differing Site Condition, and this entitles Acme to an increase in the Contract Time and Contract Sum, as defined in the Delay and Damages sections of this claim. [CONCLUSION]

### Example – *CRAC Example: Type 1 Differing Site Condition Delay Narrative*

Acme Construction is entitled to a time extension of 30 calendar days as a result of the Type 1 Differing Site Condition (DSC) that Acme encountered on the Project. [CONCLUSION] Acme utilized the Time Impact Analysis (TIA) methodology to calculate this delay. This methodology is required for all time extensions per Section 5 of the Agreement, which is noted below. This methodology requires a fragnet of delay activities to be inserted into Acme's most recent schedule update before the delay occurred. Once these activities are inserted and the schedule is updated, the impact to the substantial completion date of the Project is calculated. [RULE]

Acme used its June 15th schedule update to insert the differing site condition activities; this is Acme's most recent schedule update before the delay started on June 20th. The unimpacted June 15th schedule is attached herein. The fragnet of delay activities that Acme inserted into this schedule include the following three items: (1) halt work due to DSC and notice Owner – 1 day; (2) procure a blasting subcontractor – 10 days; and (3) blast and excavate rock – 19 days. Each of these activities has a finish-start relationship with no lag. Backup for these items including the blasting subcontractor's agreement and daily time sheets are attached to this claim. In order to properly insert these activities into the June 15th schedule, Acme broke down its total 30-day duration foundation activity into two activities—foundation excavation before the DSC (12 days) and foundation excavation after the DSC (8 days). Acme selected these durations as Acme worked on the foundation excavation for 12 days before it encountered the rock outcropping. When the fragnet of activities is inserted into the schedule and then statused, the result is a 30-day impact to the substantial completion date of the project. This day-for-day delay is understandable as the foundation work is on the critical path of the Project. A copy of this impacted schedule is attached herein. [ANALYSIS]

In sum, the rock outcropping differing site condition caused a 30-calendar day impact to the Project schedule based on Acme's TIA for this issue. Because the delay impacted the critical path of the Project, the delay is excusable and compensable. Accordingly, Acme is entitled to a 30-day time extension as well as direct costs associated with this work and extended general conditions for this additional time period, as calculated in the damage section of this claim. [CONCLUSION]

**Example – *CRAC Example: Type 1 Differing Site Condition Damages Narrative***

Acme Construction is entitled to a $75,000 increase in the Contract Sum due to direct costs associated with the Differing Site Condition work as well as extended general conditions for a 30-day time period. [CONCLUSION]. Inorder to calculate and present the damages associated with this claim, Acme utilized the industry-preferred actual cost method to present the blasting and rock excavation costs and utilized the contractually agreed upon daily rate for extended general conditions. [RULE]

Acme's earthwork and concrete subcontractor on the project is XYZ Excavation. XYZ Excavation solicited bids from three different specialty blasting companies to perform the necessary blasting work and it ultimately contracted with the lowest bidder, Pioneer Inc. Copies of the bids as well as XYZ Excavation's contract with Pioneer Inc. are attached.

After XYZ Excavation's allowable market (15% per Acme's subcontract with XYZ), as well as Acme's allowable market (5% per Section 9 of the Agreement), the total direct costs for this work is $50,000. In terms of extended general conditions, the contract provides for $833/day of excusable and compensable delay. Because the delay for the Project's substantial completion date is 30 days, Acme is entitled to extended general conditions in the amount of $25,000, which brings the total damage amount up to $75,000. [ANALYSIS]

In sum, Acme's damages related to this differing site condition claim is $75,000 based on actual costs associated with the additional work as well as 30 days of extended general conditions pursuant to the rate stipulated in the Agreement. [CONCLUSION]

# 9

# Step 8: Non-Binding Dispute Resolution

## I. Introduction

Typical dispute resolution provisions of agreements related to a construction project involve mediation and then the selected form of binding dispute resolution—litigation or arbitration. The ConsensusDocs family of contracts includes settlement meetings as a precursor to mediation. Certain public agencies also mandate that claims be presented to dispute resolution boards. Whatever the procedure may be, claimants and respondents should make good faith efforts to resolve disputes in each phase of the process. Moreover, claimants and respondents should attempt to present facts and positions in a calm and objective manner. This chapter of the book deals with how claimants and respondents should conduct themselves in this process. Both parties should again realize that claims can take months, if not years, to work through the dispute resolution process so taking this into consideration during the process is important.

## II. Prevailing Party Provisions

Before claimants proceed to binding dispute resolution, it is important to identify whether the contract agreement between the parties includes a prevailing party fee provision, where attorneys' fees are awarded to the winner of the dispute. Standard AIA and EJCDC documents do not provide for attorneys' fees. However, standard ConsensusDocs provide that the "prevailing" party gets reasonable attorneys' fees. Standard contract forms are always edited before parties enter into the agreement, so it is important to investigate whether these standard provisions were revised during the contract formation process.

One question that often arises when prevailing party provisions are incorporated into a contract is: who is the prevailing party? Courts and arbitrators vary in their treatment of these clauses—some allow the recovery of fees to the party that substantially prevailed on its claims, while others may apportion legal fees based on the percentage of victory. Prevailing party provisions often have a large impact on claimed damages. If a claimant has a $250k claim against the respondent before the binding dispute resolution phase, this amount might double if the claim has to be arbitrated or litigated because of the legal fees incurred through this process. Thus, the parties should consider this fact during non-binding settlement discussions.

## III. Settlement Meetings

All too often, claimants use claim negotiation meetings as an opportunity to vent their anger and frustration by yelling, pounding on tables, and making demands. Some respondents, on the other hand, attend settlement meetings as a means to get to binding dispute resolution and refuse to look at the issue in dispute in an objective fashion. When this is the case, this phase of dispute resolution can be a waste of time and can even be detrimental to downstream recovery as some settlement meetings are not confidential and reckless or dismissive conduct can be evidence of unprofessional and bad faith conduct.

In certain instances, it may be beneficial for the parties to engage in confidential settlement meetings under the applicable rule, which allows the parties to speak freely and make offers without fear that the offers will be disclosed during future dispute resolution proceedings. Settlement meetings can be held under Rule 408 of the Federal Rules of Evidence that allows for such candor. All 50 states have similar rules (often Rule 408 as well). Typically, settlement meetings are most effective when conducted in confidentiality.

Settlement conferences are sometimes mandated by the courts. These conferences with the court are best conducted after it is clear that key issues are preventing the parties from coming to an agreement on their own. Formal settlement meetings like this often take place shortly before trial.

It is important to act courteously, be civil, and be professional during settlement meetings. Because these meetings are typically off the record, there is no reason to put on a show. Moreover, it is important not to interrupt the opposing party. If you extend this courtesy to the opposing party, they will likely extend this courtesy back. It is also important to have a plan to demonstrate to the respondent that the claim has merits, or for the respondent to demonstrate why the claim lacks merit, based on the process noted above.

Also, most claims are not perfect and there are often two sides to every story, so understanding the opposing party's position is critical so proper rebuttals can be addressed in a timely manner. It is also important to listen and take good notes during this process so you can later recall the opposing party's position to prepare for mediation or binding dispute resolution.

# IV.  Mediation

Counsel is not required at mediation, but typically they are present at mediations. When claimants represent themselves at mediation, it is called *pro se*. Contractors should proceed with non-binding mediation and good faith. Construction projects often produce disputes. Mediation can help resolve such disputes, quickly, confidentially, and efficiently. This chapter offers tips for success in mediation.

## A.  Mediation Venue

Standard contract forms often stipulate mediation venues (i.e., AAA, JAMS, JAG, etc.) and the process for selecting a mediator. The claimant typically triggers the mediation process by filing for mediation at one of these venues, which can be done online at larger mediation venues. While the venues can simply appoint an impartial mediator, it is best to select a mediator who is familiar with construction disputes and is mutually respected by the parties. Parties typically agree to conduct mediation in a confidential setting, so discussions and information presented at mediations cannot be used against the presenting party at a later date.

## B.  Mediation Statements

Parties often agree to issue mediation statements to the mediator prior to mediation so the mediator has a better understanding of the dispute and each party's position from a high level. The parties typically agree to a page limit to mediation statements, which is often set at five or ten pages.

## C. Mediation Format

Mediations can be conducted in a joint session and then breakout sessions, or just breakout sessions. Joint sessions include the claimant and respondent. After the mediator introduces himself or herself and lays out the ground rules, the attorney for each party typically issues an opening statement and then, in some instances, the parties and/or their expert representatives provide a formal presentation to the mediator and the opposing party or parties. It is common for joint sessions to be counterproductive as parties can get defensive and emotional during joint sessions. However, it can also be important for executives from each party to hear the other side's story as issues are often more complicated than they seem.

After joint sessions, parties typically relocate into breakout rooms for separate sessions with the mediator. Initial breakout sessions may be lengthy as the mediator often gets a better understanding of each party's position and their thoughts regarding settlement. As the day progresses, the mediator moves back and forth between the breakout rooms and attempts to get a settlement range. Initial offers can be insultingly low to the opposing party but each party should know this is simply part of the process. Later in the day it becomes evident whether the parties will eventually close in on a settlement number. If the matter is not settled, additional mediations are sometimes scheduled. It is common for disputes to be settled in the second or third mediations.

## D. Settlement Paperwork

If settlement is reached at the mediation, the parties should enter into a binding term sheet that outlines the essential terms of the agreement. Typically, counsel for one party is tasked with drafting the term sheet for the other party's review. Leaving mediations with just a handshake agreement is discouraged. Note that it is common for both parties to be somewhat discouraged at the conclusion of a mediation, but that is often the sign of a good settlement.

## E. Learn from the Process

Claimants and respondents can learn from mediations, particularly on what each party needs to improve upon in the future. In addition, each party generally gets a feel for what the opposing party focuses on during binding dispute resolution. This information is valuable and typically identifies the weaknesses in each of the parties' case. Ignoring the other party's position is akin to looking at a gift horse in the mouth. The parties need to remember that dispute resolution is a process, and neither party has complete control.

## F.   Mediation Presentations

The parties often make a presentation during joint and/or breakout sessions. Presentation materials, such as PowerPoints, should not be verbose with words and should focus on screenshots of key documents, photographs, and summary statements. Presentations are often best when kept to 30 minutes or less. Similar to claim narratives, claim presentations should be objective and factual and should run through the entitlement, delay, and damages components of the claim with citations to industry standard methodologies. When in joint sessions, the opposing party will be evaluating how well the presenter will likely handle herself or himself during deposition testimony or trial/arbitration testimony, so this may weigh into whether or not a case settles at mediation. Note that: likeable and positive people who can withstand cross-examination typically make good witnesses at trial or arbitration.

## G.   The Mediation Process Is Purposefully Exhausting

Mediation is often an all-day endeavor, and sometimes an all-day and all-night adventure. The process is slow as there is significant down-time and waiting involved. Moreover, many disputes involve multiple mediations—particularly larger disputes. The back and forth between the parties, and the waiting, actually serve a purpose. The more worn out the parties, the more likely settlement will take place. Also, the more the claimant and the respondent hear about the other party's position—generally through the mediator—the more empathy and understanding will develop, and the parties generally come to the conclusion that their respective positions are not a "slam dunk." The key for claimants and respondents is to understand and respect the process. Failure to appreciate the process typically leads to unsuccessful mediations.

## 10

## Step 9: Binding Dispute Resolution

Binding dispute resolution formally settles the dispute. It takes place via litigation or arbitration depending on the venue agreed upon by the parties under the contract or based on proper jurisdiction. This section reviews all the phases and aspects of binding dispute resolution.

## I.  Litigation

Construction disputes that proceed to binding dispute resolution in the form of litigation involve civil lawsuits, as opposed to criminal lawsuits. Civil lawsuits generally proceed through distinct steps, including pleadings, discovery, trial, and possibly an appeal. Parties may file certain motions before, during, and after the trial. Note that parties can terminate the litigation by voluntarily settling at any time. For civil lawsuits, the standard of proof is by "the preponderance of evidence." The finder of fact of a litigated matter is either a judge or jury. The party filing the claim is called the plaintiff and the party who the claim is against is called the defendant. While parties have the right to represent themselves in court and file a lawsuit without an attorney (known as *pro se*), this is unusual. The following is a brief description of the steps of a typical civil trial in chronological order.

- **Venue Selection:** Contract agreements typically specify the venue for litigation. In some disputes the question arises of whether the matter should be handled in state court or federal court. If the dispute involves a federal construction projects, then federal court has jurisdiction to hear the dispute. Also, if there is complete diversity amongst the parties (the parties are headquartered in different states) and the dispute amount exceeds $75,000, the case may be brought in federal court even if the case does not involve a federal question—note that cases that fit this profile may often be brought in state court as well. In matters that do not involve federal issues, state courts have jurisdiction to hear nearly all other disputes.

- **Pleadings:** Each party in a lawsuit files initial "pleadings," which explain each party's side of the dispute. Litigated construction disputes start when one party files a complaint against the other party. Common pleadings include:
  - **Complaint:** Litigation begins when the plaintiff (claimant) files a complaint with the court and formally delivers a copy to the defendant (respondent). The complaint describes what the defendant did or failed to do that caused harm to the plaintiff and the legal basis for holding the defendant responsible for that harm.
  - **The Answer:** The defendant is given a specific amount of time to file an answer to the complaint. The answer provides the defendant's side of the dispute. The defendant may also file counterclaims against the plaintiff, alleging that the plaintiff has harmed the defendant and should be held liable for that harm. The plaintiff may respond to the defendant's answer or counterclaims by filing a reply.
- **Case Management Order:** The parties and the judge for the case develop a case management order that sets forth deadlines for the case that relate to issues such as:
  - Deadline for Exchanging Preliminary Witnesses and Exhibit Lists
  - Fact Discovery Cut-Off Date
  - Deadlines for Filing Motions to Amend and/or to Add Parties
  - Deadline for Plaintiff's Expert Designations
  - Deadline for Defendant's Expert Designations
  - Deadline for Rebuttal of Expert Designations
  - Dispositive Motion Filing Deadline
  - Pretrial Conference Date
  - Settlement Conference Date
- **Discovery:** Discovery is the process where relevant information is exchanged between the parties and third parties. Discovery is usually the longest part of the case. It generally begins soon after the complaint is answered and often does not stop until shortly before trial. In addition to producing documentation and other information, the parties may

ask each other questions in a written format known as interrogatories and requests for admission or an oral format via depositions. If a party fails to produce information, other parties my file a motion to compel, which asks the court to enforce a request for information relevant to a case.

- **Motions:** Before trial, the parties may use motions to ask the court to rule or act. Motions usually pertain to law or facts in the case, but sometimes they seek clarification or resolution of procedural disputes between the parties. Some motions, such as a motion for summary judgment, ask the court to dismiss part or all of a plaintiff's case or a defendant's defense, or dispose of issues without trial. Other motions might ask the court to order a party to produce documents or to exclude evidence from trial.

- **Briefs:** Shortly before the trial both parties issue a brief to the judge that outlines the arguments and evidence to be used at trial. In certain courts, trial briefs may not be required for jury trials.

- **Voir Dire:** Some trials do not involve a jury and are decided by the judge—these are called "bench trials." Other trials are jury trials. In a jury trial, both parties question potential jurors during a selection process known as "voir dire." Civil trials generally have between 6 to 12 jurors, depending on the venue.

- **Trial:** Trials typically start when the plaintiff gives its opening statement to the judge or jury, which is followed by the defendant's opening statement. Thereafter, the plaintiff presents its case-in-chief, where the plaintiff presents its evidence and calls witnesses to the stand for testimony. This is followed by the defendant's case-in-chief. The party that calls a witness takes direct examination of the witness and then the opposing party may cross-examine the witness, which may be followed by redirect examination and then recross-examination, if necessary. Sometimes, a party may present rebuttal evidence. During the trial a party might file and argue a trial motion to the court, such as motions in limine, motions for mistrial, and motions for judgment as a matter of law. After each party completes its case-in-chief, the plaintiff and then the defendant make closing arguments. A closing argument is the party's chance to argue to the judge or jury how and why the facts and evidence presented during the trial support a verdict in its favor. After closing arguments, and in the event of a jury trial, the court instructs the jury on the law to be applied to the evidence.

- **Verdict:** The jury or judge deliberates the case and makes a decision in favor of the plaintiff or the defendant, determining liability and the amount of money damages.
- **Post-Verdict:** A party may challenge a jury's verdict for reasons such as a trial court's errors of law or a jury's disregard of law or evidence. A motion for judgment notwithstanding the verdict asks the court to disregard the jury's verdict and enter a different decision. A motion for a new trial asks the court to set aside the jury's verdict and order a new trial of the case.
- **Appeal:** Following trial, a party dissatisfied with the lower court's decision may appeal to an appellate court. The parties present their arguments in briefs along with a record of evidence from the trial court. The appellate court will either affirm the verdict or find an error. If an error is found, the appellate court may reverse the verdict or order a new trial. Note that appellate courts will typically not review factual evidence or override a jury's findings of fact.

The duration of a civil lawsuit depends on the issues of the case, the amount of discovery to be conducted, and court scheduling and availability. The parties, guided by the rules of court, usually decide the timing of discovery while trial dates are set by the court. Litigated matters that involve construction disputes are rarely settled within one year and often go on for years. As for the duration of the trial for a typical construction dispute, it will generally last one to four weeks.

## II. Arbitration

Arbitration is a form of binding dispute resolution that is less formal than litigation in terms of the rules of evidence and other procedure. The party filing for arbitration is the claimant and the party that the claim is made against is the respondent. The dispute is decided by an arbitrator or panel of arbitrators, depending on the contract terms or mutual decisions of the parties. An arbitrator is typically an expert in the subject matter of the dispute and has had formal training in arbitration. Many, but not all, arbitrators are lawyers. Many arbitrators are also retired judges. Thus, the dispute is settled by the arbitrator or arbitrators, not by a judge or jury. Perhaps the main difference between arbitration and litigation is that arbitration awards are very difficult to appeal, and appeals are very rarely made. Hence, the arbitration award is for all intents and purposes a final decision, whereas lower court awards may be appealed to higher courts. The following is list of steps and components of a typical arbitration:

- **Filing of the Demand:** The claimant files a demand for arbitration with an appropriate arbitration venue and within the time period, if any, noted within the subject contract between the parties. The two most common arbitration venues are AAA (American Arbitration Association) and JAMS (Judicial Arbitration and Mediation Services). Both AAA and JAMS allow for online filing. Post filing, the arbitration venue typically alerts the parties that the filing requirements have been satisfied.
- **Respondent Files Answer and Counterclaims:** Within a specific time after the filing for arbitration, the respondent may file an answering statement and assert counterclaims. If no answer is filed, the respondent is deemed to deny the claim.
- **Arbitrator Appointment:** If the parties cannot agree on an arbitrator, the arbitration venue will have procedural rules on how to reach an appointment, which often includes the venue making the appointment itself based upon the facts of the case. Note that most arbitrations are heard and determined by one arbitrator. However, on certain matters (generally larger cases), it is common for each party to name one arbitrator and those two arbitrators then appoint a chairperson. Ex parte communications with an appointed arbitrator are generally prohibited.
- **Representation:** Any party may participate without representation (*pro se*), or by counsel. The majority of parties are represented by counsel.
- **Preliminary Management Hearing:** This hearing often covers issues that include, but are not limited to, the following:
  - whether all parties are included in the arbitration;
  - which arbitration rules will govern;
  - which procedural and substantive law will govern;
  - whether any dispositive issues exist;
  - whether there should be a consolidation with another arbitration;
  - whether the proceedings should be bifurcated;
  - whether mediation should be scheduled;
  - review of document production and search parameters;
  - whether the parties intend calling expert witnesses and, if so, whether to establish a schedule for the identification of experts, and identify the subject matter of their anticipated testimonies;
  - whether the arbitration will be in person or virtual;

- whether there will be a stenographic transcript or other record of the proceeding;
- whether any procedure needs to be established for the issuance of subpoenas;
- identification of any ongoing, related litigation or arbitration;
- whether post-hearing submissions will be filed;
- the form of the award.
- whether there will be site visits; and
- any other matters important to the parties or arbitrator.

- **Document Production:** The arbitrator typically manages the exchange of documents between the parties including documents that the parties intend to rely upon as well as documents the other party reasonably requests and believes will be relevant and material to the outcome of disputed issues.

- **Oaths:** Arbitrators typically require witnesses to testify under oath administered by any duly qualified person.

- **Stenographic Record:** A party that desires a stenographic to record the arbitration typically makes arrangements directly with a stenographer and will notify the other parties of these arrangements. The parties typically split these costs.

- **Arbitration Proceedings:** Parties typically agree to submit arbitration briefs to the arbitrator. Arbitrations may or may not have opening statements. The claimant then presents evidence to support its claim. The respondent then presents evidence supporting its defense in a similar manner. Oral closing statements are often waived in lieu of closing briefs, which are filed within a stipulated time after the hearing. Called witnesses are subject to direct examination, cross-examination, as well as questions from the arbitrator. The arbitrator typically has discretion to vary this procedure, provided that the parties are treated with equality.

- **Dispositive Motions:** Before, during, or after the arbitration and before the award, parties are often allowed to file motions that dispose of all or part of a claim, or narrow the issues in a case.

- **Evidence:** The parties may offer evidence that is relevant and material to the dispute. Conformity to legal rules of evidence is generally not necessary. The arbitrator determines the admissibility, relevance, and materiality of the evidence offered. The arbitrator may request offers of proof and may reject evidence deemed by the arbitrator to be cumulative, unreliable, or unnecessary.

- **Inspection or Investigation:** An arbitrator may find it necessary to make a site inspection or other investigation in connection with the arbitration and will coordinate with the parties regarding schedule—typically all parties must be present.

- **Closing of Hearing and Time of Award:** After the parties have no further proof to offer or have no further witnesses to call, the arbitrator typically declares the hearing closed unless closing briefs are to be filed, in which case the hearing is deemed to be closed upon receipt of the briefs. The time limit in which the arbitrator is required to make the award is generally 30 days from the close of the hearing, unless otherwise agreed upon.

- **Majority Decision:** When the matter is to be decided by a panel of arbitrators, a majority of the arbitrators is required to make decisions; however, if all parties and all arbitrators agree, the chair of the panel may make procedural decisions.

- **Form of Award:** Awards are made in writing and are signed by the arbitrator. Typically, the arbitrator provides a financial breakdown of any monetary award. If the parties disagree with respect to the form of the award, the arbitrator then determines the form to use.

- **Modification of Award:** Within an agreed-upon time period after the award, any party may request that the arbitrator correct any clerical or computational errors in the award. The opposing party is then given time to respond to such a request. The arbitrator then decides on the request within a short time thereafter.

The duration of an arbitration depends on the issues of the case, the amount of discovery to be conducted, and the arbitrator's schedule. The parties, guided by the rules of the arbitration, usually decide the timing of discovery and the arbitration dates are worked out by the parties and the arbitrator. Arbitrated matters that involve construction disputes, on average, are slightly faster than litigated matters but they can be considered comparable in terms of timing for all intents and purposes, so most arbitrations last from one to three weeks.

## III. Discovery and Disclosures

FRCP (Federal Rules of Civil Procedure) Rule 26 covers general disclosure requirements for federal cases. State rules regarding discovery and disclosures are largely modeled after FRCP Rule 26. Rule 26 requires the parties to produce initial disclosures that: (1) list individuals likely to have discoverable information; (2) include all information that may be used to support its claims and defenses; (3) compute each category of damages that the disclosing party intends on making along with associated backup; and (4) include insurance agreements that might satisfy all or part of a judgment. The time for initial disclosures is typically agreed to by the parties.

In addition, Rule 26 mandates that the parties disclose the names of all the expert witnesses it intends to call at trial and this disclosure must be accompanied by an expert report which must include all the opinions that it plans to express at trial, along with the basis for each opinion. The expert report must also include a listing of the information considered by the witness in forming his or her opinions. Lastly, the report must include: (1) the witness's qualifications and all publications authored in the previous 10 years; (2) a listing of cases in which the witness has testified as an expert over the last four years; and (3) a statement of the compensation to be paid for the report and for testimony in the case. If the court does not require expert reports, the disclosure must list the opinions that the witness is expected to testify.

---

**FRCP 26. Duty to Disclose; General Provisions Governing Discovery**

a) Required Disclosures.
  1) Initial Disclosure.
    A) In General. Except as exempted by Rule 26(a)(1)(B) or as otherwise stipulated or ordered by the court, a party must, without awaiting a discovery request, provide to the other parties:
      i) the name and, if known, the address and telephone number of each individual likely to have discoverable information—along with the subjects of that information—that the disclosing party may use to support its claims or defenses, unless the use would be solely for impeachment;

    ii) a copy—or a description by category and location—of all documents, electronically stored information, and tangible things that the disclosing party has in its possession, custody, or control and may use to support its claims or defenses, unless the use would be solely for impeachment;

   iii) a computation of each category of damages claimed by the disclosing party—who must also make available for inspection and copying as under Rule 34 the documents or other evidentiary material, unless privileged or protected from disclosure, on which each computation is based, including materials bearing on the nature and extent of injuries suffered; and

   iv) for inspection and copying as under Rule 34, any insurance agreement under which an insurance business may be liable to satisfy all or part of a possible judgment in the action or to indemnify or reimburse for payments made to satisfy the judgment.

B) Proceedings Exempt from Initial Disclosure. The following proceedings are exempt from initial disclosure:

    i) an action for review on an administrative record;

    ii) a forfeiture action in rem arising from a federal statute;

   iii) a petition for habeas corpus or any other proceeding to challenge a criminal conviction or sentence;

   iv) an action brought without an attorney by a person in the custody of the United States, a state, or a state subdivision;

    v) an action to enforce or quash an administrative summons or subpoena;

   vi) an action by the United States to recover benefit payments;

   vii) an action by the United States to collect on a student loan guaranteed by the United States;

  viii) a proceeding ancillary to a proceeding in another court; and

   ix) an action to enforce an arbitration award.

C) Time for Initial Disclosures—In General. A party must make the initial disclosures at or within 14 days after the parties' Rule 26(f) conference unless a different time is set by stipulation or court order, or unless a party objects during the conference that initial disclosures are not appropriate in this action and states the objection in the proposed discovery plan. In ruling on the objection, the court must determine what disclosures, if any, are to be made and must set the time for disclosure.

D) Time for Initial Disclosures—For Parties Served or Joined Later. A party that is first served or otherwise joined after the Rule 26(f) conference must make the initial disclosures within 30 days after being served or joined, unless a different time is set by stipulation or court order.

E) Basis for Initial Disclosure; Unacceptable Excuses. A party must make its initial disclosures based on the information then reasonably available to it. A party is not excused from making its disclosures because it has not fully investigated the case or because it challenges the sufficiency of another party's disclosures or because another party has not made its disclosures.

2) Disclosure of Expert Testimony.

A) In General. In addition to the disclosures required by Rule 26(a)(1), a party must disclose to the other parties the identity of any witness it may use at trial to present evidence under Federal Rule of Evidence 702, 703, or 705.

B) Witnesses Who Must Provide a Written Report. Unless otherwise stipulated or ordered by the court, this disclosure must be accompanied by a written report—prepared and signed by the witness—if the witness is one retained or specially employed to provide expert testimony in the case or one whose duties as the party's employee regularly involve giving expert testimony. The report must contain:

i) a complete statement of all opinions the witness will express and the basis and reasons for them;

ii) the facts or data considered by the witness in forming them;

iii) any exhibits that will be used to summarize or support them;

iv) the witness's qualifications, including a list of all publications authored in the previous 10 years;

v) a list of all other cases in which, during the previous 4 years, the witness testified as an expert at trial or by deposition; and

vi) a statement of the compensation to be paid for the study and testimony in the case.

C) Witnesses Who Do Not Provide a Written Report. Unless otherwise stipulated or ordered by the court, if the witness is not required to provide a written report, this disclosure must state:

i) the subject matter on which the witness is expected to present evidence under Federal Rule of Evidence 702, 703, or 705; and

ii) a summary of the facts and opinions to which the witness is expected to testify.

D) Time to Disclose Expert Testimony. A party must make these disclosures at the times and in the sequence that the court orders. Absent a stipulation or a court order, the disclosures must be made:

i) at least 90 days before the date set for trial or for the case to be ready for trial; or

ii) if the evidence is intended solely to contradict or rebut evidence on the same subject matter identified by another party under Rule 26(a)(2)(B) or (C), within 30 days after the other party's disclosure.

E) Supplementing the Disclosure. The parties must supplement these disclosures when required under Rule 26(e).

---

## IV. Witness Testimony

Witnesses testify under oath during depositions, trials, and arbitration hearings. Fact witnesses have personal knowledge of the issues surrounding the dispute that is the subject of the underlying lawsuit or arbitration. Anyone may testify as to facts of a case, but only expert witnesses may present opinions. Fact witnesses are usually laypersons with little testifying experience. Parties often engage expert witnesses to help explain technical information that often validates an argument. In a construction dispute, experts are often used to prove damages, evaluate the standard of care of a party, review liability issues, conduct forensic schedule analyses, or evaluate entitlement issues.

### A. When Can Expert Testimony Be Used?

Experts can testify in litigation when the expert's testimony will assist the trier of fact, whether the trier is a judge, jury, or arbitrator(s). Perhaps the most cited test for determining whether expert testimony should be allowable was written by Mason Ladd in his 1952 *Vanderbilt Law Review* article:

> There is no more certain test for determining when experts may be used than the common sense inquiry whether the untrained layman would be qualified to determine intelligently and to the best possible degree the particular issue without enlightenment from those having a specialized understanding of the subject involved in the dispute. (Ladd, Expert Testimony, 5 *Vand. L. Rev.* 414, 418 (1952))

In general, the exclusion of expert testimony is the exception not the norm. Judges and arbitrators often allow the introduction of expert testimony and then weigh it accordingly.

## B. Requirements for Expert Testimony

Per Federal Rule of Evidence, Rule 702, "Testimony by Expert Witnesses," expert witnesses must have "knowledge, skill, experience, training, or education" that will "help the trier of fact to understand the evidence or to determine a fact in issue." This is a broad standard. However, effective expert witnesses that meet this standard are typically objective, independent, credible, and good communicators. Thus, having the expert experience is only half the battle.

---

**FRE 702. Testimony by Expert Witnesses**  A witness who is qualified as an expert by knowledge, skill, experience, training, or education may testify in the form of an opinion or otherwise if:

a) the expert's scientific, technical, or other specialized knowledge will help the trier of fact to understand the evidence or to determine a fact in issue;
b) the testimony is based on sufficient facts or data;
c) the testimony is the product of reliable principles and methods; and
d) the expert has reliably applied the principles and methods to the facts of the case.

---

Faced with a proffer of expert scientific testimony under Rule 702, the trial judge will not focus on the expert witness' conclusions, but on his or her qualifications, principles, and methodology.

Federal Rules of Evidence, Rule 703 generally allows expert witnesses to testify based on otherwise inadmissible evidence such as hearsay opinions of other experts or non-experts. For instance, expert witnesses may rely on research performed by their own support teams as well as conversations and work product of fact witnesses.

---

**Federal Rules of Evidence, Rule 703, Bases of an Expert**  An expert may base an opinion on facts or data in the case that the expert has been made aware of or personally observed. If experts in the particular field would reasonably rely on those kinds of facts or data in forming an opinion on the subject, they need not be admissible for the opinion to be admitted. But if the facts or data would otherwise be inadmissible, the proponent of the opinion may disclose them to the jury only if their probative value in helping the jury evaluate the opinion substantially outweighs their prejudicial effect.

---

The admissibility of expert testimony is based on the Federal Rules of Evidence, Rule 104(a), which notes the court must decide any preliminary questions regarding whether a witness is qualified. The party offering the expert has the burden of establishing that the expert is properly qualified by a preponderance of evidence. See *Bourjaily v. United States*, 483 U.S. 171 (1987).

---

**Federal Rules of Evidence, Rule 104, Preliminary Questions**

a) In General. The court must decide any preliminary question about whether a witness is qualified, a privilege exists, or evidence is admissible. In so deciding, the court is not bound by evidence rules, except those on privilege.

---

## C.  How Parties Can Exclude Expert Witness Testimony

In federal matters, a party may make a Daubert motion in an attempt to exclude the opposing party's expert testimony. These motions are a specific type of motion in limine, which is discussed outside the presence of a jury and is decided by a judge. The name Daubert is based on the 1993 US Supreme Court case *Daubert, et al. v. Merrell Dow Pharmaceuticals, Inc.*, *509 U.S. 579* (1993), where the Court charged trial judges to act as gatekeepers to exclude unreliable expert testimony. While the Daubert case specifically discussed *scientific* expert testimony, the Court later clarified its position in the *Kumho Tire Co. v. Carmichael (119 S.Ct. 1167 (1999))* decision, which holds that a trial judge is the gatekeeper for all, not just scientific, expert testimony.

The Daubert decision noted several uncodified factors that a judge should consider in determining whether expert opinions are admissible. The Kumho decision notes that not all Daubert factors can be applied to every type of expert testimony. See 119 S. Ct. at 1175. These Daubert factors include:

- whether the expert's technique or theory can be or has been tested. in other words, can the expert's theory be challenged in some objective sense, or whether it is instead simply a subjective, conclusory approach that cannot reasonably be assessed for reliability;
- whether the technique or theory has been subject to peer review and publication;
- whether there is a known or potential rate of error of the technique or theory when applied;
- whether there is the existence and maintenance of standards and controls;
- whether the technique or theory has been generally accepted in the scientific community.

While Daubert is a federal court case, most states have adopted Daubert, and other states have adopted modified versions of Daubert. The states with modified Daubert versions include Alaska, California, Colorado, Connecticut, Georgia, Hawaii, Idaho, Indiana, Iowa, Maine, Maryland, Montana, New Mexico, Tennessee, Texas, Utah, and West Virginia.

## V. Deposition Testimony

A deposition is a discovery tool where the examining attorney asks the witness questions, and the witness testifies under oath that the testimony will be true and correct.[1] Persons involved in a deposition include:

- **Deponent:** The witness who provides sworn testimony at a deposition.
- **Examining Attorney:** The attorney who requests and takes the witness's deposition.

- **Defending Attorney:** The attorney who attends the deposition with the witness and objects, as necessary, to the examining attorney's questions.
- **Court Reporter:** The person recording the deposition. The court reporter also swears in the deponent prior to his or her testimony. The court reporter issues a printed transcript after the deposition. Deponents are often asked to review the transcript and to sign an affidavit that attests to the accuracy of the transcript.

## A. Deposition Rules

FRCP Rule 30, "Depositions by Oral Examination," provides the rules for when a deposition may be taken, the required notice of deposition, the rules associated with the deposition examination, edits to the deposition transcript, and penalties for failure to attend a deposition or serve a subpoena. Most states have adopted a slightly modified version of this rule. Key provisions include:

1) parties can depose any person or party;
2) the deponent must be given reasonable notice;
3) the court reporter should place the deponent under oath;
4) objections shall be placed on the record but the testimony shall proceed subject to the objection;
5) unless stipulated otherwise, depositions may last one to seven hours;
6) the deponent or a party may move to terminate or limit the deposition if it feels the deposition is conducted in bad faith or in a manner that unreasonably annoys, embarrasses, or oppresses the deponent or party;
7) the deponent has 30 days to review and make changes to the transcript;
8) the court reporter must certify the deposition.

---

### FRCP, Rule 30. Depositions by Oral Examination

a) When a deposition may be taken. A party may, by oral question, depose any person or party and the deponent's attendance may be compelled by subpoena.
b) Notice of the deposition; other formal requirements. The deponent must be given reasonable written notice, which dates the time and place of the deposition. The deponent has to produce documents if it is served a subpoena duces tecum that lists the materials designated for production. The party that notices the deposition must also state the method for recording the testimony. The notice or subpoena may

be directed to an organization and, if so, the named organization must then designate one or more persons that can testify upon its behalf.

c) Examination and Cross-examination; record of the examination; objections; written questions. The court reporter shall place the deponent under oath and record the testimony. Objections must be made on the record but the examination shall proceed--the testimony is just subject to any objection. A person may instruct a deponent not to answer only when necessary to preserve a privilege.

d) Duration; motion to terminate or limit. Unless otherwise stipulated or ordered by the court, a deposition is limited to 1 day of 7 hours. The court must allow additional time if needed to fairly examine the deponent or if the deponent, another person, or any other circumstance impedes or delays the examination. *At any time during a deposition, the deponent or a party may move to terminate or limit it on the ground that it is being conducted in bad faith or in a manner that unreasonably annoys, embarrasses, or oppresses the deponent or party* [emphasis added]. The motion may be filed in the court where the action is pending or the deposition is being taken. If the objecting deponent or party so demands, the deposition must be suspended for the time necessary to obtain an order.

e) Review by the witness; changes. The deponent must be allowed 30 days after being notified by the officer that the transcript or recording is available in which to review the transcript or recording; and if there are changes in form or substance, to sign a statement listing the changes and the reasons for making them.

f) Certification and delivery; exhibits; copies of the transcript or recording; filing. The court reporter must certify in writing that the witness was duly sworn in and that the deposition accurately records the witness's testimony. The certificate must accompany the record of the deposition. Documents and tangible things produced for inspection during a deposition must, on a party's request, be marked for identification and attached to the deposition.

g) Failure to attend a deposition or serve a subpoena; expenses: A party who, expecting a deposition to be taken, attends in person or by an attorney may recover reasonable expenses for attending, including attorney's fees, if the noticing party failed to attend and proceed with the deposition; or serve a subpoena on a nonparty deponent, who consequently did not attend.

## B. Typical Deposition Testimony Process

The process is fairly consistent from deposition to deposition, so for those who have never been a deponent, is helpful to understand the common steps of a typical deposition:

1) The examining attorney issues the deponent a notice of deposition and possibly a subpoena duces tecum that lists the information that the deponent is to bring to the deposition.

2) The deponent discusses the logistics and anticipated line of questioning with the defending attorney at least one day before the deposition is scheduled.

3) At the deposition, the deponent typically sits across from the examining attorney and next to the defending attorney and court reporter.

4) The deposition may start when the court reporter swears in the deponent typically by asking him or her to raise their right hand and asking if the deponent will tell the truth and nothing but the truth.

5) The examining attorney will generally start by reviewing some deposition ground rules with the deponent such as the need for the deponent to give verbal answers, speak up, ask for clarification when needed, and feel free to take necessary comfort breaks after he or she finishes answering the pending question. Note that the examining attorney(s) may ask the deponent about any non-privileged discussions that the deponent has during breaks regarding the case.

6) Next, the examining attorney typically asks about the deponent's past experience and current job title and duties.

7) The examining attorney will often then ask the deponent probing questions related to the matter and during this process the examining attorney may show the deponent marked exhibits or will mark new exhibits and will hand them to the deponent for review.

8) If the deponent is an expert witness for the opposing party, the examining attorney typically reviews the deponent's expert report and asks about the basis for certain opinions and asks if the deponent considered other information, such as recent deposition testimony.

9) Once the examining attorney is done asking questions, he or she will "pass the witness" or note that he or she has "no further questions." The examining attorney might note that it reserves the right to ask further questions based on further testimony. If other parties are present at the deposition, there might be questions from other examining attorneys.

10) After all questions from the examining attorney or attorneys, the defending attorney might ask the deponent questions to clarify earlier testimony. After the defending attorney is complete with his or her line of questions, the examining attorney has a chance to ask the deponent about the defending attorney's questions and related testimony. Note that the questions from the defending attorney and any subsequent questions from the examining attorney typically do not take much time.

11) The deposition then concludes, and the deponent is free to leave. Within a couple of days of the deposition, the court reporter issues the deposition transcript to the attorneys and the deponent for review and correction, if necessary, and in some instances the signature of the deponent.

## C. Recommendations for Deposition Testimony

Depositions are nerve-wracking for most deponents because it is outside most people's comfort zone. Here are some tips for deponents:

- Maintain your composure.
- Be respectful.
- Be truthful.
- Be confident.
- Be objective.
- Be independent.
- For deponents who are expert witnesses, be sure to understand the basis for all opinions listed in expert reports.
- Listen carefully to the question and answer what is asked.
- Allow the defending attorney to make objections before answering any questions.

## D. Be Mindful of the Following Scenarios During Depositions

Examining attorneys may use the following tactics to throw deponents off guard during the course of a deposition:

- **Rapid fire questioning:** If the examining attorney starts asking rapid fire questions, it can sometimes create a situation where the deponent has little time to think. Deponents should be cognizant of that and not fall for issuing rapid fire answers. Purposely pace your responses in this scenario.

- **Hypothetical questions:** Often an examining attorney will pose hypothetical questions to the deponent that have very little detail. For instance, if the deponent is asked, "Assume that the building envelope was not built per code, wouldn't the contractor be responsible rectifying this condition?" While this question may sound basic, it is riddled with traps. To properly answer such a question, you would need to: review the contractor's agreement with the owner; know what code violation exists; consider whether a reasonable similarly situation contractor might be aware of such a code provision; evaluate the design of the building envelope; figure out if the contractor's scope of work included this work; and understand if the contractors subcontracted this scope of work to another party. That said, it is not improper for the deponent to note that it cannot properly answer such a hypothetical question without additional information.

- **Rabbit holes:** If a deponent answers a question even though he or she is not completely sure of the answer, the examining attorney might drill down on this issue and either expose the deponent's uncertainty or try to get the deponent to answer further questions despite the lack of clarity If a deponent finds himself or herself in a rabbit hole, the best route to take at that point is to clean up the previous answer and note that you are unsure about the issue and that answering further questions on this subject would be speculative and that the line of questioning has brought to light uncertainty.

- **Demand for a yes or no answer:** Examining attorneys often ask questions and look for a "yes" or "no" answer from the deponent; however, many questions cannot be properly answered with a "yes" or "no," and often require answers to be qualified, such as:
  - "Not necessarily…"
  - "It depends…"
  - "I would need to know more…"
  - "Yes (or no), because…"
  - "A 'yes' or 'no' answer would be deceptive to the reader of my transcript because…"

For instance, if an examining attorney asks a deponent a question and is simply looking for a "yes" or "no" answer, he or she might ask, "If the contractor did not properly flash a window, and in fact you actually see the light between the window frame and the adjacent wall when you stand on the inside of the building, this will lead to water infiltration, correct?" The proper answer to this question is not "yes" or "no," it would be, "Not necessarily, because there might be a large soffit above the window that will prevent moisture from getting to the window location." Deponents need to be ready for situations where the examining attorneys demands a "yes" or "no" answer: "please, just answer 'yes' or 'no,' it is simple question." Some deponents get worn down by these questions and deliver a "yes" or "no" answer even when the proper answer is otherwise.

- **Yes to Death:** Examining attorneys sometimes ask a series of questions that require simple a simple "yes" answer and this lulls the deponent into "yes" answers and then adds some questions where a "yes" answer may not be appropriate.
- **Best Friend Examining Attorneys:** Some examining attorneys can be extremely cordial and likable during depositions. While this often puts the deponent at ease, it can also cause deponents to put their guard down and not listen for tricky questions.
- **Bully Examining Attorneys:** Other examining attorneys use bully tactics to intimidate the deponent and this can lead deponents to give the examining attorney the answer he or she is looking for rather than cause an uncomfortable back and forth exchange.
- **Not Showing the Document:** Examining attorneys may ask the deponent specific questions about a document without showing the deponent the document. If the deponent either doesn't recall the answer or can't answer the question without reviewing the document, the deponent should note that it can't answer the question without reviewing the document.
- **Embarrassing the Deponent for Not Knowing the Answer:** Construction projects typically generate thousands of documents related to plans, specifications, contracts, RFIs, change orders, ASIs, daily reports, inspection results, etc. and deponents involved in the subject project can't be expected to know all this information without reviewing the actual documents. Examining attorneys might ask specific questions regarding certain documents and may appear extremely surprised if the deponent cannot answer the question off the top of his or her head, and this can be embarrassing for a deponent. If the deponent at one point knew the answer to the question but doesn't know it when the question is posed, a proper answer is, "I don't recall ..." If the deponent never knew the answer to this question, a proper answer is "I don't know ..."

## E. Deposition Testimony Used as Impeachment at Trial

Cross-examining attorneys often try to impeach witnesses at trial or arbitration with deposition testimony previously given by witnesses. Thus, it is important for witnesses to review their deposition transcripts before trial or arbitration because cross-examining attorneys are often off base when they try to impeach witnesses by taking deposition testimony out of context. For instance, consider this line of questioning:

CROSS-EXAMINING ATTORNEY:   Mr. Smith, isn't it true that your company installed flooring material on top of the concrete floors on the project?

WITNESS:   Yes.

CROSS-EXAMINING ATTORNEY:   The subcontract agreement between your flooring firm and the general contractor notes that your firms is responsible for covering up defective work, correct?

WITNESS:   Yes.

CROSS-EXAMINING ATTORNEY:   Because the concrete floors are defective, your firm covered up defective work, correct?

WITNESS:   No.

CROSS-EXAMINING ATTORNEY:   Mr. Smith, you gave deposition testimony a month ago on this project, right?

WITNESS:   Yes.

CROSS-EXAMINING ATTORNEY:   You were under oath during that deposition, similar to your testimony today, and you agreed to tell the truth and nothing but the truth, right?

WITNESS:   Yes.

CROSS-EXAMINING ATTORNEY:   On page 112 of your deposition transcript, starting on line 4, you testified that, "I agree the concrete floor slabs at the project are defective," isn't that true?

WITNESS:   Yes, but if you continue reading my deposition testimony, I go on to clarify that my firm did not know that the concrete floor slabs were defective at the time of flooring installation as the issue related to a latent concrete mix issue that was not discovered until well after the flooring work was complete.

As noted in the example above, the witness properly recalled his deposition testimony and avoided impeachment. If the witness had not recalled his deposition testimony, the examining attorney might have successfully impeached the witness.

## VI.   Trial and Arbitration Testimony

The process of witness testimony at trial is similar to that at arbitration. Attorneys call witnesses to the stand in an orderly fashion and witnesses get sworn in by judges, arbitrators, or court reporters before any testimony is given. Fact witnesses are not qualified by judges or arbitrators, but experts are. Testimony for experts generally starts with a review of relevant qualifications and the attorney taking the expert's direct examination will proffer the expert to the judge or arbitrator as an expert in one or more subject matters. Opposing counsel may challenge or voir dire the witness's qualifications before the judge or arbitrator makes a ruling or opposing counsel can "stipulate" that the witness is a qualified expert. The judge or arbitrator will then qualify or reject the witness as an expert.

On direct examination, the attorney taking the direct examination may not lead the witness. An example of this would be, "Isn't it true that you believe the defendant is responsible for the improper window installation at the project?" A more appropriate and open-ended line of questioning would be, "Did you review and investigate the window installation at the Project?" Witness answers, "Yes." "Based on this investigation, what are your opinions regarding the installation?" On cross-examination, the attorney taking the cross-examination may lead the witness. After cross-examination, the attorney who took the direct examination may redirect the witness but the line of questioning is limited to topics that were discussed during the cross-examination. Thereafter, the opposing counsel my recross-examine the witness.

The recommendations for deposition testimony are the same recommendations for trial or arbitration testimony, which include maintaining your composure, being respectful, being truthful, being confident, being objective, being independent, being prepared, and being a good listener. In addition, when one of the attorneys makes an objection to the judge or arbitrator, the witness must pause and allow the judge to rule on the objection. If the judge or arbitrator sustains the objection of the opposing counsel, the witness may not answer the question. If the judge or arbitrator overrules the objection, the witness may answer the question.

At the conclusion of the witness's testimony, the attorneys for the plaintiff and defendant will note that they have no further questions. After this, the judge or arbitrator will formally release the witness, at which time the witness may step down from the witness stand and exit the courtroom or arbitration venue. Some other procedural requirements for witnesses are to always rise when the judge or jury enters and exits the courtroom. The judge will indicate when the witness (and all other participants) may be seated. In addition, if a witness is later called in for rebuttal testimony, the judge will remind the witness that he or she remains under oath for the testimony, so there is no need to re-swear in the witness. In certain matters, the parties might agree to sequester non-party witnesses, which means these witnesses can only be in the courthouse or arbitration when he or she is testifying.

Persons involved in a trial or arbitration include:

- **Judge:** The elected or appointed person who has the power to make decisions on cases brought before a court of law. In bench trials, judges decide the prevailing party and the amount of the award.
- **Jury:** A body of persons legally selected and sworn to inquire into any matter of fact and to give their verdict according to the evidence.
- **Arbitrator(s):** The person who arbitrates the dispute and makes decisions on issues regarding the dispute.
- **Party Representatives:** The person or persons who are present during the litigation or arbitration and sit at the plaintiff/claimant's or defendant/respondent's table.
- **Attorney(s) for the Plaintiff or Claimant:** The attorney or attorneys who represent the plaintiff/claimant, who sit at the plaintiff's/claimant's table.
- **Attorney(s) for the Defendant or Respondent:** The attorney or attorneys who represent the defendant/respondent who sit at the defendant/respondent's table.
- **Fact Witnesses:** Lay witnesses who are called to discuss facts surrounding the dispute.
- **Expert Witnesses:** Witnesses who are called to discuss expert opinions regarding the dispute.
- **Court Reporter:** The person recording the trial or arbitration. The court reporter may swear in witnesses before his or her testimony.

## Note

1 See https://civilprocedure.uslegal.com/discovery/discovery-devices/

## 11

## Termination Claims

Most construction contracts provide termination provisions where the paying party can terminate the performing party for convenience or for cause, and the performing party can terminate the paying party for cause. Termination for cause is the most severe remedy that one party has against the other, so it is imperative that terminations are procedurally correct and have substantive basis. This chapter reviews each of these termination scenarios. For the purpose of this chapter, the parties include an owner and its contractor and the agreement between the parties is a standard AIA contract form.

## I. Termination for Convenience (Owner Termination of Contractor for Convenience)

Most contracts allow the owner to terminate the contract agreement for the owner's convenience and without cause. If such a provision exists, the owner's decision to cancel the contract for convenience is not a breach under the contract. If a termination for convenience clause is not included in the contract, an owner election to quit the contract would constitute a breach.

A termination for convenience is often referred to as a "T for C." In this scenario, it is rare that owner-contractor disputes arise regarding the propriety of the T for C itself, as contract language is typically clear that the owner can terminate the contract for convenience at any time and without any stated reason. The amount owed to the contractor upon a T for C, however, is often disputed. An owner typically elects to T for C a contract agreement if the contract no longer makes financial sense for the owner or if the owner has lost faith in the contractor even though the contractor has not substantially breached the terms of the contract.

## A. Calculation of the Final Payment Due to the Contractor Under a T for C

Most contracts note that in the event of a termination for convenience, the owner shall pay the contractor for work properly performed, plus termination-related expenses, plus a termination fee (if defined per contract), and minus previous owner payments to the contractor. In lieu of a defined termination fee provision, some contracts allow the contractor to make a claim for lost profits on unexecuted work. Contractors typically calculate amounts owed after a T for C through a schedule of values analysis or a modified total cost claim analysis.

### 1. Calculation of the Value of Work Performed as of the T for C

A schedule of values analysis is most easily done with unit price contracts as the parties can generally agree upon the quantities of work completed as of the date of termination. For non-unit price line items or non-unit price contracts, the evaluation is more challenging, particularly because divergent subcontractor and vendor positions often exist regarding the percentage complete for various line items within the schedule of values. The use of photographs and digital quantity takeoffs is a good way to establish the amount of installed quantities. For partially installed work, such as a sidewalk that is graded and formed but has not been poured, the contractor has the burden of valuing this work and incorporating it into the percentage complete of the line item. In the sidewalk example, if 50% of the sidewalks are placed and 20% of the sidewalks were formed and ready for pour, it might be reasonable to bring the overall total to 60% complete.

Contractors often utilize the modified total cost method to determine the reasonable value of the installed work, particularly when the termination for convenience occurs before the work is 50% complete, as the schedule of values method can sometimes not account for all early mobilization charges, particularly if an average general conditions amount per month is used in the contractor's payment applications. If such a methodology is selected, the contractor simply needs to produce its job cost report and associated backup, and then apply the as-bid overhead and profit percentage to the direct costs of the work. If the owner disagrees with the contractor's stated position regarding the value of work that it has completed, the contractor can perform both analyses to support its position.

### 2. Should the Contractor Account for Defective Work?

Another issue that is ordinarily not worked out per the terms of the T for C clause is how to account for defective work—should the cost to correct the work be reduced from the overall earned value or should the work not be considered in the percentage complete? The AIA A201 notes that the contractor is entitled to work "properly" performed in the event of a T for C, so this is clear that the contractor would not be entitled to costs related to defective work. On the other hand, the T for C language in the ConsensusDocs 200 simply notes that contractor is entitled to recover for work "performed to date," so this leaves the question of defective work unanswered. Although this issue is not always defined in the terms of the contract, it is reasonable to assume that contractor should offer some form of discount to account for defective work. If the defective work is not a contractor responsibility, for instance, if the defect is a result of an owner design issue, no deduction is necessary. To wit, if the owner under-designed the steel girders on a bridge project and the girders excessively deflected after placement of the bridge deck, the owner would be responsible for this issue, and the contractor would not be obligated to offer a deduct in its T for C claim.

### 3. Termination-Related Costs

In addition to the calculating the value of work performed by the contractor, most T for C provisions allow for the recovery of termination-related costs. Such costs may include legal fees related to the termination, administrative and management fees to manage the T for C process, outside consulting fees to assist with tasks such as the calculation of the value of work performed, penalties related to the cancelation of subcontractor and vendor agreements, demobilization charges, etc. These fees should be itemized and supported with actual cost backup to the extent possible.

### 4. Lost Profits or Termination Fees or Neither

T for C provisions in owner-contractor agreements either: (1) preclude the recovery of lost profits on unexecuted work; (2) are silent on the issue; (3) allow for a lost profit claim on unexecuted work; or (4) contemplate a termination fee to be paid to contractor.

Over the past two decades, standard contractor forms, such as the AIA, have moved away from allowing the contractor to recover the cost for lost profits on unexecuted work to allowing a termination fee. Updated ConsensusDocs forms have also included a termination fee. EJCDC forms and the FAR explicitly preclude the recovery of lost profits. Many proprietary contract forms continue to allow the contractor to recover for lost profits in the event of a T for C, and many owners continue to use the 2007 version of the A201 general conditions, and under this version, lost profits are recoverable. When contract forms describe allowable costs in the event of a T for C, but do not explicitly preclude the recovery of lost profits, it is typical to assume that lost profits are not recoverable due to a lack of inclusion. However, if a contract does not describe which costs are recoverable under a T for C, lost profits are generally included in T for C claims.

The calculation of lost profit on cost-plus contracts differs slightly from that of stipulated sum contracts. For instance, if the owner-contractor agreement is a CM at-risk form, the contractor fee is a separate line item that is calculated based on a negotiated percentage that is applied to the cost of the work. If the balance of unearned cost of work is $55M, and the negotiated fee is 4.25% of the cost of the work, the lost profit would be calculated by multiplying the 4.25% times the $55M to get to $2.475M in lost profit.

If the owner-contractor agreement is a stipulated sum in nature, the contractor must demonstrate its anticipated profit percentage for the project. This is done by comparing the original contract amount with its original estimate for the work. Once this profit percentage is identified, the lost profit is then calculated. For example, if the contractor completed 72% of the work on a $20M project per the schedule of values, the remaining value of work is 18% of $20M, or $3.6M ($20M x 18%). If the contractor's anticipated profit of the work was 5%, and the unearned work is $3.6M, then you don't simply multiply 5% times $3.6M, as this would calculate profit on top of profit ($180k). Rather, the equation is:

Lost Profit = Unearned Work – (Unearned Work / (1 + Profit%).

Hence, the calculation is:

$3.6M – ($3.6M/1.05) = Lost Profit of **$171, 428.57**

Thus, the contractor can claim $171,428.57 for lost profit in its T for C proposal.

### 5. Subtraction of Past Owner Payments to Contractor

The last step in calculating a T for C claim amount is to total up the allowable categories of costs and subtract all prior payments the owner has made to the contractor to determine the contractor's net T for C recovery amount.

**Example – *T for C Claim***

The owner and the contractor entered a $130M stipulated sum contract for the construction of a high-rise office building. The contract indicates that the owner can T for C the contractor at any time, and if this occurs, the contractor is entitled to the value of the work properly executed, plus termination-related costs, plus a termination fee of $500k, minus all owner payments to contractor.

During the course of construction, the owner elects to T for C the contractor after it completes 72% of the work, which is based on the contractor's final schedule of values analysis. Thus, 72% x $130M = $93.6M. In addition, the contractor incurred $852k in termination fees, which is based on four items: (1) legal fees related to the cancelation of all subcontractor and vendor contracts (supported by invoices); (2) termination fees paid to subcontractors and vendors (supported by final account statements that note any termination fees and include full and final lien releases); (3) management fees related to the cancelation of the subcontractor and vendor payments (supported by daily reports); and (4) demobilization charges (supported by timesheets and invoices). Also, the contractor's T for C proposal includes an invoice for the $500k termination fee that is bargained for in the contract. As of the date of termination, the owner has paid the contractor $86.7M.

## T for C proposal

| | |
|---|---|
| Value of Work Properly Executed | $93,600,000 |
| Termination-Related Costs | $852,000 |
| Termination Fee | $500,000 |
| Paid To Date | $(86,700,000) |
| Net Amount Due Contractor | $8,252,000 |

As shown above, the contractor is entitled to a final T for C payment of $8.252M. In the event that the owner disputes this amount, the parties can move forward with dispute resolution to address the differences.

### B.   T for C Provisions in Standard Contract Forms

**AIA A201: Section 14.4.3** In case of such termination for the Owner's convenience, the Owner shall pay the Contractor for Work properly executed; costs incurred by reason of the termination, including costs attributable to termination of Subcontracts; and the termination fee, if any, set forth in the Agreement.

**ConsensusDocs 200: Section 11.4.2** If Owner terminates this Agreement for convenience, Constructor shall be paid: (a) for the Work performed to date including Overhead and profit; (b) for all demobilization costs and costs incurred resulting from termination, but not including Overhead or profit on Work not performed; (c) reasonable attorneys' fees and costs related to termination; and (d) a premium as follows: $XX.

**EJCDC C700: Section 16.03** Owner May Terminate for Convenience

A. Upon 7 days' written notice to Contractor and Engineer, Owner may, without cause and without prejudice to any other right or remedy of Owner, terminate the Contract. In such case, Contractor shall be paid for (without duplication of any items):

1. Completed and acceptable Work executed in accordance with the Contract Documents prior to the effective date of termination, including fair and reasonable sums for overhead and profit on such work;

2. Expenses sustained prior to the effective date of termination in performing services and furnishing labor, materials, or equipment as required by the Contract Documents in connection with uncompleted Work, plus fair and reasonable sums for overhead and profit on such expenses; and

3. Other reasonable expenses directly attributable to termination, including costs incurred to prepare a termination for convenience cost proposal.

B. Contractor shall not be paid for any loss or anticipated profits or revenue, post-termination overhead costs, or other economic loss arising out of or resulting from such termination.

**FAR Clause 52.249-2:** Termination for Convenience of the Government

(f) ... Contractor and the Contracting Officer may agree upon the whole or any part of the amount to be paid or remaining to be paid because of the termination. The amount may include a reasonable allowance for profit on work done. However, the agreed amount, whether under this paragraph (f) or paragraph (g) of this clause, exclusive of costs shown in paragraph (g)(3) of this clause, may not exceed the total contract price as reduced by (1) the amount of payments previously made and (2) the contract price of work not terminated. The contract shall be modified, and the Contractor paid the agreed amount. Paragraph (g) of this clause shall not limit, restrict, or affect the amount that may be agreed upon to be paid under this paragraph.

(g) If the Contractor and the Contracting Officer fail to agree on the whole amount to be paid because of the termination of work, the Contracting Officer shall pay the Contractor the amounts determined by the Contracting Officer as follows, but without duplication of any amounts agreed on under paragraph (f) of this clause:

(1) The contract price for completed supplies or services accepted by the Government (or sold or acquired under paragraph (b)(9) of this clause) not previously paid for, adjusted for any saving of freight and other charges.

(2) The total of:

(i) The costs incurred in the performance of the work terminated, including initial costs and preparatory expense allocable thereto, but excluding any costs attributable to supplies or services paid or to be paid under paragraph (g)(1) of this clause;

(ii) The cost of settling and paying termination settlement proposals under terminated subcontracts that are properly chargeable to the terminated portion of the contract if not included in subdivision (g)(2)(i) of this clause; and

(iii) A sum, as profit on subdivision (g)(2)(i) of this clause, determined by the Contracting Officer under 49.202 of the Federal Acquisition Regulation, in effect on the date of this contract, to be fair and reasonable; however, if it appears that the Contractor would have sustained a loss on the entire contract had it been completed, the Contracting Officer shall allow no profit under this subdivision (g)(2)(iii) and shall reduce the settlement to reflect the indicated rate of loss.

(3) The reasonable costs of settlement of the work terminated, including:

(i) Accounting, legal, clerical, and other expenses reasonably necessary for the preparation of termination settlement proposals and supporting data;

     (ii) The termination and settlement of subcontracts (excluding the amounts of such settlements); and

     (iii) Storage, transportation, and other costs incurred, reasonably necessary for the preservation, protection, or disposition of the termination inventory.

(h) Except for normal spoilage, and except to the extent that the Government expressly assumed the risk of loss, the Contracting Officer shall exclude from the amounts payable to the Contractor under paragraph (g) of this clause, the fair value as determined by the Contracting Officer, for the loss of the Government property.

(i) The cost principles and procedures of part 31 of the Federal Acquisition Regulation, in effect on the date of this contract, shall govern all costs claimed, agreed to, or determined under this clause.

(j) The Contractor shall have the right of appeal, under the Disputes clause, from any determination made by the Contracting Officer under paragraph (e), (g), or (l) of this clause, except that if the Contractor failed to submit the termination settlement proposal or request for equitable adjustment within the time provided in paragraph (e) or (l), respectively, and failed to request a time extension, there is no right of appeal.

(k) In arriving at the amount due the Contractor under this clause, there shall be deducted:

     (1) All unliquidated advance or other payments to the Contractor under the terminated portion of this contract;

     (2) Any claim which the Government has against the Contractor under this contract; and

     (3) The agreed price for, or the proceeds of sale of, materials, supplies, or other things acquired by the Contractor or sold under the provisions of this clause and not recovered by or credited to the Government.

(l) If the termination is partial, the Contractor may file a proposal with the Contracting Officer for an equitable adjustment of the price(s) of the continued portion of the contract. The Contracting Officer shall make any equitable adjustment agreed upon. Any proposal by the Contractor for an equitable adjustment under this clause shall be requested within 90 days from the effective date of termination unless extended in writing by the Contracting Officer.

(m) (1) The Government may, under the terms and conditions it prescribes, make partial payments and payments against costs incurred by the Contractor for the terminated portion of the contract, if the Contracting Officer believes the total of these payments will not exceed the amount to which the Contractor will be entitled.

(2) If the total payments exceed the amount finally determined to be due, the Contractor shall repay the excess to the Government upon demand, together with interest computed at the rate established by the Secretary of the Treasury under 50 U.S.C. App.1215(b)(2). Interest shall be computed for the period from the date the excess payment is received by the Contractor to the date the excess is repaid. Interest shall not be charged on any excess payment due to a reduction in the Contractor's termination settlement proposal because of retention or other disposition of termination inventory until 10 days after the date of the retention or disposition, or a later date determined by the Contracting Officer because of the circumstances.

## II. Termination for Cause (Owner Termination of Contract with Contractor)

Contracts often detail the terminable reasons for a termination for cause, which is often referred to as a termination for default, or "T for D." In general, most contracts require the owner to provide the contractor with a formal notice of termination to the contractor so that the contractor can cure, or start to cure, the default. If the contractor does not cure the default, or take reasonable steps to start curing the default, the owner can terminate the contract for cause. Under this scenario, the contractor typically owes the owner the completion and termination-related costs that are above and beyond the contractor's remaining unpaid contract balance. In the event that the owner's T for D costs fall below the contractor's remaining unpaid contract balance, which is extremely rare, the owner owes the contractor the difference between the two.

A T for D can have a grave financial effect on the contractor. First, owner completion cost claims often far exceed the contractor's remaining contract balance. Second, a T for D can impact the contractor's ability to procure future work because owner RFPs often inquire whether the contractor has been terminated for cause in the past. Accordingly, the owner must have good reason for a T for D and must precisely follow procedural requirements. In addition, if the contract is bonded by a surety provider, and the owner seeks to demand surety performance, the owner must follow both the T for D terms of the contract as well as the terms of the performance bond. Hence, it is critical for the owner to understand and follow the procedures of both. It is also critical for the owner to define the specific reason or reasons for T for D because most contracts require the owner to provide the contractor with a cure period where the contractor can address actionable issues in order prevent a T for D. Because of the contentious nature of T for Ds, as well as the severe consequences regarding the same, the propriety of T for Ds are often disputed by contractors and surety providers.

Contract and bond forms have varying T for D procedures. For instance, Section 14.2 of the A201 notes that the owner may terminate the contractor for cause for four reasons: (1) contractor repeatedly refuses to provide enough adequate workers or materials to the project; (2) contractor fails to make proper payment to its subcontractors or suppliers; (3) contractor repeatedly fails to follow laws/statutes/ordinances/rules and regulations/code/orders of public authority; or (4) an otherwise substantial breach of a provision of the contract documents. Thus, the owner's latitude to issue a T for D to the contractor is broad, but key terms include "repeated" and "substantial."

Per Section 14.2.2 of the A201, if any of these four reasons exist and the owner seeks to issue a T for D, the owner must issue the contractor and its surety a seven-day notice, obtain certification from the architect that sufficient cause exists for a T for D, and follow the terms of the performance bond. If the owner properly follows these steps, the owner may exclude the contractor from the site, take possession of the contractor's materials, equipment, tools, construction equipment, and machinery that are on site, accept assignment of its subcontractors, and complete the work in an expedient fashion. Also, in the event of a T for D, the contractor is not entitled to any additional payments until the work is complete and if the unpaid contract balance is insufficient to cover the owner's completion costs and termination fees, which is almost always the case, the contractor owes the owner the difference. Note that the initial decision maker must certify the amount of this differential before the owner makes its claim for this amount.

If the owner seeks performance from the contractor's surety provider in the event of possible T for D, the AIA A312 performance bond notes that the surety's obligation does not arise until the owner undertakes three steps. First, Section 3.1 notes that the owner shall provide notice to the contractor and the surety that the owner is considering declaring a contractor default, and whether the owner is requesting a conference between the three parties to discuss the contractor's performance. If the owner does not request a conference, the surety may do so within five days of receipt of the owner's notice. The A312 notes that a Section 3.1 conference shall be held within ten business days after the surety's receipt of notice from the owner. Second, the owner must terminate the contract and notify the surety of such termination. Third, the owner shall agree to pledge the contractor's unpaid contract balance per the terms of the contract to the surety or to a completion contractor. Note that Section 4 states that if the owner fails to complete the notice requirement of Section 3.1, surety is still obligated under the terms of the bond except to the extent it is prejudiced by such owner failure.

If surety's obligations are properly triggered under Section 3, Section 5 provides the surety with five options. First, the surety can arrange for the terminated contractor to complete the work with consent from the owner. Second, the surety can complete the balance of work as a completing surety through an independent contractor. Third, the surety can relet the work and tender a bonded and qualified contractor to the owner along with payment of any shortfall in funds. Fourth, the surety can determine and pay an amount it determines it is liable to the owner. Or, fifth, it can deny liability in whole or part and notify the owner of its reasons for denial.

## A. T for D Provisions in Standard Contract Forms (Termination by Owner)

### AIA A201 (2017): §14.2 Termination by the Owner for Cause

§14.2.1 The Owner may terminate the Contract if the Contractor

1. repeatedly refuses or fails to supply enough properly skilled workers or proper materials;
2. fails to make payment to Subcontractors or suppliers in accordance with the respective agreements between the Contractor and the Subcontractors or suppliers;

      3. repeatedly disregards applicable laws, statutes, ordinances, codes, rules and regulations, or lawful orders of a public authority; or

      4. otherwise is guilty of substantial breach of a provision of the Contract Documents.

§14.2.2  When any of the reasons described in Section 14.2.1 exist, and upon certification by the Architect that sufficient cause exists to justify such action, the Owner may, without prejudice to any other rights or remedies of the Owner and after giving the Contractor and the Contractor's surety, if any, seven days' notice, terminate employment of the Contractor and may, subject to any prior rights of the surety:

      1. Exclude the Contractor from the site and take possession of all materials, equipment, tools, and construction equipment and machinery thereon owned by the Contractor;

      2. Accept assignment of subcontracts pursuant to Section 5.4; and

      3. Finish the Work by whatever reasonable method the Owner may deem expedient. Upon written request of the Contractor, the Owner shall furnish to the Contractor a detailed accounting of the costs incurred by the Owner in finishing the Work.

§14.2.3  When the Owner terminates the Contract for one of the reasons stated in Section 14.2.1, the Contractor shall not be entitled to receive further payment until the Work is finished.

§14.2.4  If the unpaid balance of the Contract Sum exceeds costs of finishing the Work, including compensation for the Architect's services and expenses made necessary thereby, and other damages incurred by the Owner and not expressly waived, such excess shall be paid to the Contractor. If such costs and damages exceed the unpaid balance, the Contractor shall pay the difference to the Owner. The amount to be paid to the Contractor or Owner, as the case may be, shall be certified by the Initial Decision Maker, upon application, and this obligation for payment shall survive termination of the Contract.

**ConsensusDocs 200 (2019): 11.3 Owner's Right to Terminate for Default**

11.3.1 Termination by Owner for Default: Upon expiration of the second notice period to cure pursuant to Section 11.2 and absent appropriate corrective action, Owner may terminate this Agreement by written notice. Termination for default is in addition to any other remedies available to Owner under Section 11.2. If Owner's costs arising out the Contractor's failure to cure, including the costs of completing the Work and reasonable attorneys' fees, exceed the unpaid Contract Price, Constructor shall be liable to Owner for such excess costs. If Owner's costs are less than the unpaid Contract Price, Owner shall pay the difference to Constructor. If Owner exercises its rights under this section, upon the request of Constructor, Owner shall furnish to Constructor a detailed accounting of the costs incurred by Owner.

11.3.2 Use of Contractor's Materials, Suppliers, and Equipment: If Owner or Others perform work under Section 11.3, Owner shall have the right to take and use any materials and supplies for which Owner has paid and located at the Worksite for the purpose of completing any remaining Work. Owner and others performing work under Section 11.3 shall also have the right to use construction tools and equipment located on the Worksite and belonging to the Constructor or Subsubcontractors for the purpose of completing the remaining Work, but only after Constructor's written consent. If Owner uses Constructor's construction tools and equipment in accordance with this subsection, then Owner shall indemnify and hold harmless Constructor and applicable Subcontractors and the agents, officers, directors, and employees of each of them, from and against all claims, damages, losses, costs, and expenses, including but not limited to reasonable attorneys' fees, costs, and construction tools and equipment. Immediately upon completion of the Work, any remaining materials, supplies, or equipment not consumed or incorporated int the Work shall be returned to Constructor in substantially the same condition as when they were taken, reasonable wear and tear expected.

    11.3.3 If Constructor files a petition under the Bankruptcy Code, this Agreement shall terminate if: (a) Constructor or Constructor's trustee rejects the Agreement; (b) a default occurred and Constructor is unable to give adequate assurance of required performance; or (c) Constructor is otherwise unable to comply with the requirements for assuming this Agreement under the applicable provisions of the Bankruptcy Code.

    11.3.4 Owner shall make reasonable efforts to mitigate damages arising from Constructor default, and shall promptly invoice Constructor for all amounts due pursuant to Section 11.2 and Section 11.3.

**EJCDC C-700 (2018): 16.02 Owner May Terminate for Cause**

  A. The occurrence of any one or more of the following events will constitute a default by

    Contractor and justify termination for cause:

    1. Contractor's persistent failure to perform the Work in accordance with the Contract Documents (including, but not limited to, failure to supply sufficient skilled workers or suitable materials or equipment, or failure to adhere to the Progress Schedule);

    2. Failure of Contractor to perform or otherwise to comply with a material term of the Contract Documents;

    3. Contractor's disregard of Laws or Regulations of any public body having jurisdiction; or

    4. Contractor's repeated disregard of the authority of Owner or Engineer.

  B. If one or more of the events identified in Paragraph 16.02.A occurs, then after giving Contractor (and any surety) 10 days' written notice that Owner is considering a declaration that Contractor is in default and termination of the Contract, Owner may proceed to:

    1. declare Contractor to be in default, and give Contractor (and any surety) written notice that the Contract is terminated; and

    2. enforce the rights available to Owner under any applicable performance bond.

  C. Subject to the terms and operation of any applicable performance bond, if Owner has terminated the Contract for cause, Owner may exclude Contractor from the Site, take possession of the Work, incorporate in the Work all materials and equipment stored at the Site or for which Owner has paid Contractor but which are stored elsewhere, and complete the Work as Owner may deem expedient.

D. Owner may not proceed with termination of the Contract under Paragraph 16.02.B if Contractor within 7 days of receipt of notice of intent to terminate begins to correct its failure to perform and proceeds diligently to cure such failure.

E. If Owner proceeds as provided in Paragraph 16.02.B, Contractor shall not be entitled to receive any further payment until the Work is completed. If the unpaid balance of the Contract Price exceeds the cost to complete the Work, including all related claims, costs, losses, and damages (including but not limited to all fees and charges of engineers, architects, attorneys, and other professionals) sustained by Owner, such excess will be paid to Contractor. If the cost to complete the Work including such related claims, costs, losses, and damages exceeds such unpaid balance, Contractor shall pay the difference to Owner. Such claims, costs, losses, and damages incurred by Owner will be reviewed by Engineer as to their reasonableness and, when so approved by Engineer, incorporated in a Change Order. When exercising any rights or remedies under this paragraph, Owner shall not be required to obtain the lowest price for the Work performed.

F. Where Contractor's services have been so terminated by Owner, the termination will not affect any rights or remedies of Owner against Contractor then existing or which may thereafter accrue, or any rights or remedies of Owner against Contractor or any surety under any payment bond or performance bond. Any retention or payment of money due Contractor by Owner will not release Contractor from liability.

G. If and to the extent that Contractor has provided a performance bond under the provisions of Paragraph 6.01.A, the provisions of that bond will govern over any inconsistent provisions of Paragraphs 16.02.B and 16.02.D.

**FAR Clause 52.249-8: Default (Fixed-Price Supply and Service)**

(a) (1) The Government may, subject to paragraphs (c) and (d) of this clause, by written notice of default to the Contractor, terminate this contract in whole or in part if the Contractor fails to—

(i) Deliver the supplies or to perform the services within the time specified in this contract or any extension;

(ii) Make progress, so as to endanger performance of this contract (but see paragraph (a)(2) of this clause); or

(iii) Perform any of the other provisions of this contract (but see paragraph (a)(2) of this clause).

(a) (2) The Government's right to terminate this contract under subdivisions (a)(1)(ii) and (1)(iii) of this clause, may be exercised if the Contractor does not cure such failure within 10 days (or more if authorized in writing by the Contracting Officer) after receipt of the notice from the Contracting Officer specifying the failure.

(b) If the Government terminates this contract in whole or in part, it may acquire, under the terms and in the manner the Contracting Officer considers appropriate, supplies or services similar to those terminated, and the Contractor will be liable to the Government for any excess costs for those supplies or services. However, the Contractor shall continue the work not terminated.

(c) Except for defaults of subcontractors at any tier, the Contractor shall not be liable for any excess costs if the failure to perform the contract arises from causes beyond the control and without the fault or negligence of the Contractor. Examples of such causes include (1) acts of God or of the public enemy, (2) acts of the Government in either its sovereign or contractual capacity, (3) fires, (4) floods, (5) epidemics, (6) quarantine restrictions, (7) strikes, (8) freight embargoes, and (9) unusually severe weather. In each instance the failure to perform must be beyond the control and without the fault or negligence of the Contractor.

(d) If the failure to perform is caused by the default of a subcontractor at any tier, and if the cause of the default is beyond the control of both the Contractor and subcontractor, and without the fault or negligence of either, the Contractor shall not be liable for any excess costs for failure to perform, unless the subcontracted supplies or services were obtainable from other sources in sufficient time for the Contractor to meet the required delivery schedule.

(e) If this contract is terminated for default, the Government may require the Contractor to transfer title and deliver to the Government, as directed by the Contracting Officer, any (1) completed supplies, and (2) partially completed supplies and materials, parts, tools, dies, jigs, fixtures, plans, drawings, information, and contract rights (collectively referred to as "manufacturing materials" in this clause) that the Contractor has specifically produced or acquired for the terminated portion of this contract. Upon direction of the

Contracting Officer, the Contractor shall also protect and preserve property in its possession in which the Government has an interest.

(f) The Government shall pay contract price for completed supplies delivered and accepted. The Contractor and Contracting Officer shall agree on the amount of payment for manufacturing materials delivered and accepted and for the protection and preservation of the property. Failure to agree will be a dispute under the Disputes clause. The Government may withhold from these amounts any sum the Contracting Officer determines to be necessary to protect the Government against loss because of outstanding liens or claims of former lien holders.

(g) If, after termination, it is determined that the Contractor was not in default, or that the default was excusable, the rights and obligations of the parties shall be the same as if the termination had been issued for the convenience of the Government.

(h) The rights and remedies of the Government in this clause are in addition to any other rights and remedies provided by law or under this contract.

## B.  Key Procedural Provisions in Standard Bond Forms

### AIA A312 (2010) Performance Bond (Key Parts)

**§3** If there is no Owner Default under the Construction Contract, the Surety's obligation under this Bond shall arise after

1. the Owner first provides notice to the Contractor and the Surety that the Owner is considering declaring a Contractor Default. Such notice shall indicate whether the Owner is requesting a conference among the Owner, Contractor and Surety to discuss the Contractor's performance. If the Owner does not request a conference, the Surety may, within five (5) business days after receipt of the Owner's notice, request such a conference. If the Surety timely requests a conference, the Owner shall attend. Unless the Owner agrees otherwise, any conference requested under this Section 3.1 shall be held within ten (10) business days of the Surety's receipt of the Owner's notice. If the Owner, the Contractor and the Surety agree, the Contractor shall be allowed a reasonable time to perform the Construction Contract, but such an agreement shall not waive the Owner's right, if any, subsequently to declare a Contractor Default;

2. the Owner declares a Contractor Default, terminates the Construction Contract and notifies the Surety; and

3. the Owner has agreed to pay the Balance of the Contract Price in accordance with the terms of the Construction Contract to the Surety or to a contractor selected to perform the Construction Contract.

**§4** Failure on the part of the Owner to comply with the notice requirement in Section 3.1 shall not constitute a failure to comply with a condition precedent to the Surety's obligations, or release the Surety from its obligations, except to the extent the Surety demonstrates actual prejudice.

**§5** When the Owner has satisfied the conditions of Section 3, the Surety shall promptly and at the Surety's expense take one of the following actions:

1) Arrange for the Contractor, with the consent of the Owner, to perform and complete the Construction Contract;

2) Undertake to perform and complete the Construction Contract itself, through its agents or independent contractors;

3) Obtain bids or negotiated proposals from qualified contractors acceptable to the Owner for a contract for performance and completion of the Construction Contract, arrange for a contract to be prepared for execution by the Owner and a contractor selected with the Owner's concurrence, to be secured with performance and payment bonds executed by a qualified surety equivalent to the bonds issued on the Construction Contract, and pay to the Owner the amount of damages as described in Section 7 in excess of the Balance of the Contract Price incurred by the Owner as a result of the Contractor Default; or

4) Waive its right to perform and complete, arrange for completion, or obtain a new contractor and with reasonable promptness under the circumstances:

1. After investigation, determine the amount for which it may be liable to the Owner and, as soon as practicable after the amount is determined, make payment to the Owner; or

2. Deny liability in whole or in part and notify the Owner, citing the reasons for denial.

**ConsensusDocs260 (2011) Performance Bond (Key Parts)**

2. SURETY OBLIGATIONS If the Contractor is in default pursuant to the Contract and the Owner has declared the Contractor in default, the Surety promptly may remedy the default or shall:

a. Complete the Work, with the consent of the Owner, through the Contractor or otherwise;

b. Arrange for the completion of the Work by a Contractor acceptable to the Owner and secured by performance and payment bonds equivalent to those for the Contract issued by a qualified surety. The Surety shall make available as the Work progresses sufficient funds to pay the cost of completion of the Work less the Contract Balance up to the Bond Sum; or

c. Waive its right to complete the Work and reimburse the Owner the amount of its reasonable costs, not to exceed the Bond Sum, to complete the Work less the Contract Balance.

3. DISPUTE RESOLUTION All disputes pursuant to this Bond shall be instituted in any court of competent jurisdiction in the location in which the Project is located and shall be commenced within two years after default of the Contractor or Substantial Completion of the Work, whichever occurs first. If this provision is prohibited by law, the minimum period of limitation available to sureties in the jurisdiction shall be applicable.

**EJCDC C-610 (2013) Performance Bond (Key Parts)**

1. Contractor and Surety, jointly and severally, bind themselves, their heirs, executors, administrators, successors, and assigns to Owner for the performance of the Construction Contract, which is incorporated herein by reference.

2. If Contractor performs the Construction Contract, Surety and Contractor shall have no obligation under this Bond, except when applicable to participate in a conference as provided in Paragraph 3.

3. If there is no Owner Default under the Construction Contract, Surety's obligation under this Bond shall arise after:

3.1 Owner first provides notice to Contractor and Surety that Owner is considering declaring a Contractor Default. Such notice shall indicate whether Owner is requesting a conference among Owner, Contractor, and Surety to discuss the Contractor's performance. If Owner does not request a conference, Surety may, within five (5) business days after receipt of Owner's notice, request such a conference. If Surety timely requests a conference, Owner shall attend. Unless Owner agrees otherwise, any conference requested under this Paragraph 3.1 shall be held within ten (10) business days of Surety's receipt of Owner's notice. If Owner, Contractor, and Surety agree, Contractor shall be allowed a reasonable time to perform the Construction Contract, but such an agreement shall not waive Owner's right, if any, subsequently to declare a Contractor Default;

3.2 Owner declares a Contractor Default, terminates the Construction Contract and notifies Surety; and

3.3 Owner has agreed to pay the Balance of the Contract Price in accordance with the terms of the Construction Contract to Surety or to a contractor selected to perform the Construction Contract.

4. Failure on the part of Owner to comply with the notice requirement in Paragraph 3.1 shall not constitute a failure to comply with a condition precedent to Surety's obligations, or release Surety from its obligations, except to the extent Surety demonstrates actual prejudice.

5. When Owner has satisfied the conditions of Paragraph 3, Surety shall promptly and at Surety's expense take one of the following actions:

5.1 Arrange for Contractor, with the consent of Owner, to perform and complete the Construction Contract;

5.2 Undertake to perform and complete the Construction Contract itself, through its agents or independent contractors;

5.3 Obtain bids or negotiated proposals from qualified contractors acceptable to Owner for a contract for performance and completion of the Construction Contract, arrange for a contract to be prepared for execution by Owner and a contractor selected with Owner's concurrence, to be secured with performance and payment bonds executed by a qualified surety equivalent to the bonds issued on the Construction Contract, and pay to Owner the amount of damages as described in Paragraph 7 in excess of the Balance of the Contract Price incurred by Owner as a result of the Contractor Default; or

5.4 Waive its right to perform and complete, arrange for completion, or obtain a new contractor, and with reasonable promptness under the circumstances:

5.4.1 After investigation, determine the amount for which it may be liable to Owner and, as soon as practicable after the amount is determined, make payment to Owner; or

5.4.2 Deny liability in whole or in part and notify Owner, citing the reasons for denial.

6. If Surety does not proceed as provided in Paragraph 5 with reasonable promptness, Surety shall be deemed to be in default on this Bond seven (7) days after receipt of an additional written notice from Owner to Surety demanding that Surety perform its obligations under this Bond, and Owner shall be entitled to enforce any remedy available to Owner. If Surety proceeds as provided in Paragraph 5.4, and Owner refuses the payment or Surety has denied liability, in whole or in part, without further notice Owner shall be entitled to enforce any remedy available to Owner.

7. If Surety elects to act under Paragraph 5.1, 5.2, or 5.3, then the responsibilities of Surety to Owner shall not be greater than those of Contractor under the Construction Contract, and the responsibilities of Owner to Surety shall not be greater than those of Owner under the Construction Contract. Subject to the commitment by Owner to pay the Balance of the Contract Price, Surety is obligated, without duplication for:

    7.1  the responsibilities of Contractor for correction of defective work and completion of the Construction Contract;

    7.2  additional legal, design professional, and delay costs resulting from the Contractor's Default, and resulting from the actions or failure to act of Surety under Paragraph 5; and

    7.3  liquidated damages, or if no liquidated damages are specified in the Construction Contract, actual damages caused by delayed performance or non-performance of Contractor.

8. If Surety elects to act under Paragraph 5.1, 5.3, or 5.4, Surety's liability is limited to the amount of this Bond.

9. Surety shall not be liable to Owner or others for obligations of Contractor that are unrelated to the Construction Contract, and the Balance of the Contract Price shall not be reduced or set off on account of any such unrelated obligations. No right of action shall accrue on this Bond to any person or entity other than Owner or its heirs, executors, administrators, successors, and assigns.

10. Surety hereby waives notice of any change, including changes of time, to the Construction Contract or to related subcontracts, purchase orders, and other obligations.

11. Any proceeding, legal or equitable, under this Bond may be instituted in any court of competent jurisdiction in the location in which the work or part of the work is located and shall be instituted within

two (2) years after a declaration of Contractor Default or within two (2) years after Contractor ceased working or within two (2) years after Surety refuses or fails to perform its obligations under this Bond, whichever occurs first. If the provisions of this paragraph are void or prohibited by law, the minimum periods of limitations available to sureties as a defense in the jurisdiction of the suit shall be applicable.

12. Notice to Surety, Owner, or Contractor shall be mailed or delivered to the address shown on the page on which their signature appears.

**FAR/GSA Performance Bond Standard Form 25A (2016) (Key Parts)**

OBLIGATION:

We, the Principal and Surety(ies), are firmly bound to the United States of America (hereinafter called the Government) in the above penal sum. For payment of the penal sum, we bind ourselves, our heirs, executors, administrators, and successors, jointly and severally. However, where the Sureties are corporations acting as co-sureties, we, the Sureties, bind ourselves in such sum "jointly and severally" as well as "severally" only for the purpose of allowing a joint action or actions against any or all of us. For all other purposes, each Surety binds itself, jointly and severally with the Principal, for the payment of the sum shown opposite the name of the Surety. If no limit of liability is indicated, the limit of liability is the full amount of the penal sum.

CONDITIONS:

The Principal has entered into the contract identified above.

THEREFORE:

The above obligation is void if the Principal-

(a) (1) Performs and fulfills all the understanding, covenants, terms, conditions, and agreements of the contract during the original term of the contract and any extensions thereof that are granted by the Government, with or without notice of the Surety(ies) and during the life of any guaranty required under the contract, and (2) Performs and fulfills all the undertakings, covenants, terms, conditions, and agreements of any and all duly authorized modifications of the contract that hereafter are made. Notice of those modifications to the Surety(ies) are waived.

(b) Pays to the Government the full amount of the taxes imposed by the Government, if the said contract is subject to 41 USC Chapter 31, Subchapter III, Bonds, which are collected, deducted, or withheld from wages paid by the Principal in carrying out the construction contract with respect to which this bond is furnished.

## III. Termination for Cause (Contractor Termination of Contract with Owner)

Most construction contracts also allow the contractor to terminate the owner for cause in certain scenarios. A contractor's termination of an owner is considerably rarer than an owner termination of a contractor, but it happens from time to time, and similar disputes often result from such draconian action. Contracts vary on what is recoverable under a contractor T for D of an owner, but it often matches, at a minimum, that of a T for C.

Section 14.1 of the A201 allows the contractor to terminate the owner, after the contractor issues a seven-day cure notice to the owner and the architect, if one of the following seven conditions exists: (1) a court order stops work for 30 or more consecutive days; (2) an act of government such as a national emergency stops work for 30 or more consecutive days; (3) the architect has not issued certificate of payment and has not notified contractor for any reason for withholding certification stops work for 30 or more consecutive days; (4) the owner has not made payment per contract stops work for 30 or more consecutive days; (5) the owner has not furnished contractor proof of funding per Section 2.2 stops work for 30 or more consecutive days; (6) if the owner suspends the work per Section 14.3 for a period that, in aggregate, exceeds the original project duration or 120 days in any 365-day period, whichever is less; or (7) if the owner causes the work to stop for a period of 60 or more consecutive days due to repeated failures by owner to fulfill its obligations under the contract with respect to matters important to the progress of work.

If one of these seven conditions exists and the contractor provided proper notice to the owner and the architect, and such issue was not properly cured, Section 14.1.3 allows the contractor to T for D the owner and recover payment for work executed, profit on work not executed, and costs resultant from the termination. Hence, the A201 allows contractor to recover for lost profits on incomplete work.

## A. T for D Provisions in Standard Contract Forms (Termination by Contractor)

### AIA A201 (2017): §14.1 Termination by the Contractor

§14.1.1 The Contractor may terminate the Contract if the Work is stopped for a period of 30 consecutive days through no act or fault of the Contractor, a Subcontractor, a Sub-subcontractor, their agents or employees, or any other persons or entities performing portions of the Work, for any of the following reasons:

1. Issuance of an order of a court or other public authority having jurisdiction that requires all Work to be stopped;
2. An act of government, such as a declaration of national emergency, that requires all Work to be stopped;
3. Because the Architect has not issued a Certificate for Payment and has not notified the Contractor of the reason for withholding certification as provided in Section 9.4.1, or because the Owner has not made payment on a Certificate for Payment within the time stated in the Contract Documents; or
4. The Owner has failed to furnish to the Contractor reasonable evidence as required by Section 2.2.

§14.1.2 The Contractor may terminate the Contract if, through no act or fault of the Contractor, a Subcontractor, a Sub-subcontractor, their agents or employees, or any other persons or entities performing portions of the Work, repeated suspensions, delays, or interruptions of the entire Work by the Owner as described in Section 14.3, constitute in the aggregate more than 100 percent of the total number of days scheduled for completion, or 120 days in any 365-day period, whichever is less.

§14.1.3 If one of the reasons described in Section 14.1.1 or 14.1.2 exists, the Contractor may, upon seven days' notice to the Owner and Architect, terminate the Contract and recover from the Owner payment for Work executed, as well as reasonable overhead and profit on Work not executed, and costs incurred by reason of such termination.

§14.1.4 If the Work is stopped for a period of 60 consecutive days through no act or fault of the Contractor, a Subcontractor, a Sub-subcontractor, or their agents or employees or any other persons or entities performing portions of the Work because the Owner has repeatedly failed to fulfill the Owner's obligations under the Contract Documents with respect to matters important to the progress of the Work, the Contractor may, upon seven additional days' notice to the Owner and the Architect, terminate the Contract and recover from the Owner as provided in Section 14.1.3.

**ConsensusDocs 200 (2019): 11.5 Constructor's Right to Terminate**

11.5.1 Seven (7) Days' after Owner's receipt of written notice from Consturctor, Constructor may terminate this Agreement if the Work as been stopped for a thirty (30) Day period thorugh no fault of Constructor for any of the following reasons:

(a) under court order or order of other governmental authorities having jurisdiction;

(b) as a result of the declaration of a national emergency or other governmental act during which, through no fact or fault of Constructor, materials are not available; or

(c) suspension by Owner for convenience pursant to Section 11.1.

11.5.2 In addition, upon seven (7) Days' written notice to Owner and an opportunity to cure within three (3) Days, Constructor may terminate this Agreement if Owner:

11.5.2.1 fails to furnish reasonable evidence pursuant to Section 4.2 that sufficient funds are available and committed for Project financing; or

11.5.2.2 assigns this Agreement over Contractor's reasonable objection; or

11.5.2.3 fails to pay Constructor in accordance with this Agreement and Constructor has stopped Work in compliance with Section 9.5; or

11.5.2.4 otherwise materially breaches this Agreement.

11.5.3 Upon termination by Constructor in accordance with Section 11.5.2, Constructor is entitled to recover from Owner payment for all Work executed and for any proven loss, cost, or expense in connection with the Work, including all demobilization costs plus reasonable Overhead and profit on Work not performed.

**EJCDC C-700 (2018): 16.04 Contractor May Stop Work or Terminate**

A. If, through no act or fault of Contractor, (1) the Work is suspended for more than 90 consecutive days by Owner or under an order of court or other public authority, or (2) Engineer fails to act on any Application for Payment within 30 days after it is submitted, or (3) Owner fails for 30 days to pay Contractor any sum finally determined to be due, then Contractor may, upon 7 days' written notice to Owner and Engineer, and provided Owner or Engineer do not remedy such suspension or failure within that time, terminate the contract and recover from Owner payment on the same terms as provided in Paragraph 16.03.

B. In lieu of terminating the Contract and without prejudice to any other right or remedy, if Engineer has failed to act on an Application for Payment within 30 days after it is submitted, or Owner has failed for 30 days to pay Contractor any sum finally determined to be due, Contractor may, 7 days after written notice to Owner and Engineer, stop the Work until payment is made of all such amounts due Contractor, including interest thereon. The provisions of this paragraph are not intended to preclude Contractor from submitting a Change Proposal for an adjustment in Contract Price or Contract Times or otherwise for expenses or damage directly attributable to Contractor's stopping the Work as permitted by this paragraph.

# 12

# Non-Contract Claims and Defenses

## I. Introduction

While the focus of this book is on contract claims, the parties to construction projects should also be familiar with non-contract claims that are often asserted in construction disputes, as well as non-contract defenses that may be asserted against contract claims. These claims and defenses are primarily based on statutory law and state and federal case law, so qualified legal counsel should be retained before asserting or rebutting such positions. However, it is important for claimants and respondents to be familiar with these doctrines, even at a cursory level. Many AEC industry professionals are familiar with these buzzwords but have no idea what they mean—the aim of this chapter is to fix that.

The non-contract claims reviewed below include quantum meruit, unjust enrichment, negligence, implied warranties, mechanic's liens, and Miller Act claims. The non-contract defenses to breach of contract claims include the doctrine of estoppel, waiver, and unconscionability.

## II. Non-Contract Claims

### A. Quantum Meruit

Quantum meruit is a non-contract claim that that means "as much as one has earned" and is used to assert the reasonable damages for services provided under one of the following three scenarios: (1) when an understanding exists between the parties, but no formal contract exists; (2) the scope of work is well beyond that contemplated in a contract; or (3) the contract terms were never agreed upon. Damages for quantum meruit claims are often calculated by valuing the benefit received by the respondent at the expense of the claimant, at the time when the respondent received the value. It is important in quantum meruit claims for the claimant to establish that the parties *intended* to enter into a formal contract, or that the work falls well beyond the scope of the contract.

An example of this is if a developer and a general contractor are negotiating contract terms and the parties agree that the general contractor should commence work on a multi-family apartment project in order to conclude foundation work before winter conditions set in. If the parties later fail to conclude the contract negotiations, and the contractor stops work on the project, the contractor can bring a quantum meruit claim against the developer. Here, the work was not performed under a contract so the developer would owe the contractor a reasonable sum for work performed at the request of the developer. In order to establish the "reasonable sum," claimants need to demonstrate that the claimed amounts represent fair market value for the scope of work—this can be done per the techniques identified in the damages chapter (Chapter 13).

## B. Unjust Enrichment

Unjust enrichment claims are often confused with quantum meruit claim, but it is important to understand the difference between the two. Unjust enrichment claims are not premised upon having an understanding between the parties, they are based upon the interest of preventing one party from retaining a benefit without payment. The key here is whether the alleged receiving party has actually received a benefit. To recover on an unjust enrichment claim, the claimant typically must prove: (1) no other remedies exist; (2) the respondent received a benefit from the claimant and the claimant appreciates and is enriched by the benefit; and (3) the respondent's acceptance and appreciation of the benefit make it unjust for it not to pay for such value.

An example of an unjust enrichment claim scenario is when a subcontractor to a general contractor fails to serve timely notice of its lien claim to the owner based on the general contractor's failure to pay the subcontractor. The subcontractor here might argue that the owner maintains sufficient contract balance to backcharge the contractor so the owner would be unjustly enriched by not paying the subcontractor. Unjust enrichment looks through the eyes of the recipient of the benefit to determine if value was in fact received. For example, if the contractor went ahead and built several offices that a landlord was planning on renting out as office space, the landlord would be in receipt of a benefit. However, if a contractor performs these same improvements in a building that the landlord was planning on demolishing, there would be no value to the landlord.

## C. Negligence

Negligence is a tort, so a negligence claim is not made under a contract between parties, but in a scenario where one party purportedly owes a duty to another party and this duty was breached and caused damage to the claimant. Hence, if the elements of negligence are proven in a construction dispute, relief may be available to the claimant regardless of whether a formal contract exists between the parties or not. To prove negligence, the claimant must establish: (1) the respondent owed a duty of care to the claimant; (2) the respondent breached this duty of care; and (3) the claimant incurred damage as a result of this breach.

In a construction dispute setting, the duty of care is often evaluated by experts that opine whether or not the respondent met its required standard of care when it designed and/or constructed the subject project. Experts typically compare the respondent's conduct with that of a similarly situated, reasonable contractor or designer. These hypothetical contractors or designers typically follow contract requirements (if a contract exists), industry standards, applicable code requirements, accepted industry practice, and manufacturer's recommendations. If these requirements conflict with one another, they are often weighed by expert witnesses.

The Economic Loss Doctrine has been adopted by most jurisdictions in the US and it serves to prevent a party from seeking greater recovery in tort than under available contract remedies. Note that in certain states, the Economic Loss Rule does not preclude negligence actions in matters that involve, for example, residential construction defect cases where state case law or state statutes have established that construction and design professionals owe a duty of care to property owners, independent of any contract. Thus, it is important to understand when the Economic Loss Rule can act as a shield against negligence claims.

State statute of limitations and statute of repose laws cut off negligence claim rights if not acted upon by a specified deadline. States puts a time limit on a potential plaintiff's right to file a civil lawsuit after suffering some form of harm through these statutes. Statute of limitations typically limit negligence claims to two to three years after any injury is suffered or reasonably recognized. Statute of repose laws limit the time period for a plaintiff to discover an injury and for construction projects the clock typically starts when construction is substantially complete. State statute of repose laws typically bar construction litigation after 4 to 12 years, depending on the state. Note that states often have exceptions for statute of limitations, but statute of repose laws are often not subject to similar limitations and are often a stricter time limit on filing suit. Note that certain states, such as New York, do not have formal statute of repose law, but they have case law or other laws that limit the time period for construction litigation to be brought, which effectively serve as a statute of repose.

## D. Breach of Implied Warranty

Construction contracts often include express warranties that the constructor is bound to. Implied warranties, on the other hand, are created by law or by courts. Certain states have case law that establishes that homebuilders impliedly warrant that the houses they construct be habitable for residential use. The Supreme Court in *United States v. Spearin*, 248 U.S. 132 (1918) noted that owners that provide plans and specifications to their contractors impliedly warrant the adequacy of their plans and specifications—this is known as the Spearin Doctrine. If a contractor completes the work per the owner's plans and specifications but there is a deficiency or failure therein that the contractor could not reasonably have known about and could not reasonably identify, the owner is responsible. Note that the owner may then assert claims against the designers for partial or full responsibility for the issue.

## E. Mechanic's Liens

The intent of mechanic's liens laws is to protect suppliers, workers, and service providers that improve real property against the unjust enrichment of a property owner. Mechanic's liens are governed by statutory provisions in each of the 50 states and the District of Columbia. All lien statutes include strict procedural requirements that must be followed by lienors. Lien laws allow parties with and without privity with the property owner to file a lien on the property. Thus, it is beyond a contractual remedy. Rather than award monetary damages, a perfected lien typically creates a right to the owner's real property.

Most states limit potential lien claimants as follows: (1) claimants be within three tiers of the project owner; (2) suppliers of tangible products must be affixed or become a permanent party of the project; or (3) no mechanic's liens exist for supplier-to-supplier dealings. To enforce a mechanic's lien, the claimant must establish that it supplied services or materials that were incorporated into the job or that it was employed by the owner, construction manager, architect, engineer, contractor or subcontractor of any tier.

States require a notice of lien or claim within a strict time period after there has been a failure to make timely payment. The claimant then files or records the mechanic's lien under penalty of perjury at the applicable country office. The title to the property is then subject to the claimant's mechanic's lien and persons taking title to the property are on notice that the claimant may have rights to the property. Note that virtually all states permit a payment bond to be substituted for the right to file for a mechanic's lien.

## F. The Miller Act and the Little-Miller Act Claims

The Miller Act of 1935, which is applied under the Federal Acquisition Regulations, regulates all payments made on federal construction improvements where the federal government has a contract in place with a constructor for more than a threshold amount (currently $150,000). All 50 states have adopted "Little" Miller Acts, which relates to state construction improvements that exceed a threshold amount, which varies from state to state. The Miller Act and Little Miller Acts require the prime contractor to furnish payment and performance bonds in the amount of the face value of the contract. These bonds provide no protection for the prime contractor.

Payment bonds typically protect first- and second-tier subcontractors and first-tier suppliers. The Miller Act and Little Miller Acts mandate that the payment bond stands in the place of federal and state property, respectively, so claimants can make a claim directly against the surety provider that underwrote the bond, and they can receive payment if their claim is successful. Like lien laws, payment bonds have strict notice requirements for payment bond claims, which must be followed. Any lawsuit to enforce your bond claim rights must typically be filed within 1 year from the claimant's last furnishing of labor or materials to the federal or state construction project.

## III. Non-Contract Defenses to Breach of Contract Claims

### A. Estoppel

The doctrine of estoppel is often asserted by parties in construction disputes where the asserting party relied upon an initial representation of the other party and then the representation is changed to the detriment of the asserting party. Estoppel acts to prevent injustices that result when one company changes a position it once took after another company has relied on the earlier position. Estoppel is often asserted under two scenarios.

First, if a supplier or subcontractor reports that it has been paid in full and the general contractor or owner relies on that representation, the supplier or subcontractor may not later be permitted to claim that it has not been paid in full. In other words, the party that initially reported that it has been paid in full may be estopped from taking a position to the contrary. This fact pattern often involves signed lien waivers that acknowledge full payment. Second, if a general contractor is bidding for a project and it relies on subcontractor bids for its pricing, and then a subcontractor withdraws or substantially changes its bid after the general contractor has submitted its bid and is awarded the project, the subcontractor may be bound to its original bid under the doctrine of estoppel. In essence, the doctrine of estoppel holds a party to its word that has been detrimentally relied upon by another party.

### B. Waiver

The waiver doctrine has been asserted by contractors on federal construction projects in the past to defend against a late application of liquidated damages related to a contractor's late completion of work when "unusual circumstances" exist. Unusual circumstances in the past have included indications by the government that the completion date is no longer enforceable—this includes scenarios where the completion date has passed and there was never any mention of an assessment of liquidated damages to the contractor. Typically, contractors that have successfully used this defense in the past establish the following three elements: (1) the government did not terminate the contract well after the contractor passed the contract completion date; (2) the contractor relies on the fact that the government did not terminate to continue performance on the project; and (3) unusual circumstances exist.

## C. Unconscionability

Unconscionability is a defense against enforcement of unfair contract provisions in a breach of contract claim. Proving unconscionability is difficult, but courts might hold a provision unenforceable if both unfair bargaining during contract formation and unfair substantive terms are shown. Unfair bargaining can be demonstrated by proving an absence of choice by the disadvantaged party. Unfair substantive terms can be demonstrated by showing how the term is unreasonably favorable to the other party. If the court deems a contract provision unconscionable, it cannot be enforced so there is no breach of contract. Common contract provisions that are often challenged as unfair include overreaching indemnity provisions, liquidated damages that act as a penalty, no damages for delay clauses, and pay-if-paid clauses.

# IV.  Summary

Claimants and respondents to breach of contract claims must also be aware of non-contract claims and defenses that exist and are often cited in construction disputes. Preparation of non-contract claims, and assertion of non-contract defenses are often assembled by attorneys that coordinate with claimants and respondents to secure the necessary information to properly assert these positions. Having no familiarity with these items can handicap claimants and respondents during dispute resolution.

# 13

## Allocation of Damages

Many disputes implicate multiple parties to a construction project, which often complicates resolution of the dispute. One way to timely resolve multi-party disputes is to reasonably allocate fault among the responsible parties. Without such allocation, multi-party claims often move to costly binding dispute resolution so a judge, jury, or arbitrator can render decisions on fault and damage allocation. This chapter provides a methodology that claimants and respondents can use to properly allocate fault and damages in order to address this dilemma.

The following methodology has been used to allocate fault on many small, large, and mega construction disputes that involve multiple parties. This process includes a seven-step review of each claimed issue:

1) a review and understanding of the specific issue in dispute;
2) a review of the duties of the various parties that are often set forth in various contract agreements;
3) a determination of whether the subject issue is patently obvious or latent in nature;
4) a review of whether a party covered up patently defective work of another party;
5) a review of third-party inspector obligations;
6) a review of indemnification clauses within written contracts;
7) a determination of the potentially responsible parties and percentages of fault amongst the parties.

It is common for parties to disagree on whether or not the subject work is defective. When this is the case, it often makes sense to allocate damages under two scenarios. The first allocation scenario assumes that the claimant's allegations are correct and that the claimant's proposed remedial actions are proper. The second scenario assumes the respondent's rebuttal position is correct and the allocation of damages is based on the respondent's scope of repair work, if any. Allocation discussions often facilitate the resolution of dispute when multiple respondents are implicated in the claimant's allegations. The following sections cover a more detailed review of these seven steps, followed by examples of this methodology.

## I. Step 1 Defining the Issue in Dispute

The first step in the fault allocation process is to clearly understand the claimant's allegations in the dispute. Regarding a dispute involving a claim of defective work, this would include an evaluation of which building components are involved and the extent of any damage caused by the alleged defect, if any. Also, the question of whether the removal and replacement of the work will damage adjacent or integrated work needs to be addressed. Another key consideration is how the installation differs from the requirements of the contract documents and, if the contract documents are deficient, how the work should have been specified or designed. This generally includes a review of information such as contemporaneous correspondence and photographs of the issue, and a review of information such as contract documents, manufacturers' recommendations, industry standard publications, applicable building codes, submittals, requests for information (RFIs), etc.

## II. Step 2 Defining the Duties of the Various Parties

Numerous parties participate in the planning, design, coordination, construction, inspection, and maintenance of construction projects. When a dispute is at issue during the construction phase of a project, it is important to understand the various duties amongst all potentially responsible parties. Responsibilities of each party are typically set forth in written contract agreements and associated contract documents. If this is not the case and the agreements are oral, then the actions of the parties and invoices for the work are important to consider. Also, it is common to run into disputes where the disputed work was performed by multiple contractors, so sorting out who did what is critical. For instance, if the issue relates to framing in five buildings, and the general contractor mobilized three framers to complete the work, the subcontractor agreements, invoices, or daily reports might define which contractors worked in each area of the project.

## III.   Step 3 Is the Issue Patent or Latent in Nature?

The next step is to analyze whether the issue is a patent or latent condition. If the alleged defect is patent—a condition which a reasonably qualified general contractor's project superintendent should have been able to readily observe—then a portion of fault is often allocated to the general contractor unless applicable contract agreements exclude such liability. To the contrary, if the alleged defect is a latent condition—a condition which a reasonably qualified general contractor's project superintendent would not have reasonably been able to observe—no fault is typically allocated to the general contractor.

## IV.   Step 4 Was the Defective Work Covered Up by a Subsequent Trade?

Standard subcontract agreements typically include a boilerplate provision whereby the subcontractor must inspect the substrate upon which its work will cover to identify any patent defects in the work and, if so, the subcontractor is responsible for reporting such defects to the general contractor or else the subcontractor becomes responsible for the patently defective substrate. If this happens, a share of the responsibility is often allocated to the party that covered up the work. On the other hand, if the condition is latent, fault is not allocated to the party that covered up the work.

## V.   Step 5 Did a Third-Party Inspector Approve the Work?

Standard contract forms typically don't relieve contractors/subcontractors of their obligation to perform in accordance with the contract documents. However, in certain circumstances, installers might rely on an inspector's approval before proceeding with work. For instance, if an earthwork contractor moisturizes and compacts a sub-base, and per contract it relies exclusively on a third-party inspector retained by the owner for approval to proceed with the work after the inspector performs compaction tests and formally advises the contractor that the work is in conformance with the contract requirements, and the compaction of the sub-base is later found to be non-conforming, then a portion of the fault shifts to the inspector.

## VI.  Step 6 Do the Contracts have Indemnification Clauses?

Most standard contract forms include indemnification provisions whereby if work is found to be non-conforming, the party specifically retained to perform this work shall fully indemnify the party paying for this work. For instance, if a contractor retains a subcontractor to furnish and install all windows and doors on a project, and during the course of construction the owner hires an inspector to review the windows and the general contractor's superintendent also reviews the installation, and it turns out that the installation was faulty, the subcontractor cannot allocate fault to the general contractor because the indemnification provision makes the subcontractor fully responsible for its work, regardless of the multiple "eyes" that were laid on this work. In certain instances, if the enforceability of indemnification clauses are at issue based on venue and applicable case law, it sometimes makes sense to run an allocation report that assumes the indemnification provision is enforceable and one that assumes it is not enforceable.

## VII.  Step 7 Identify the Responsibility of the Various Parties

Once the analyst concludes steps 1 through 6, a determination can be made regarding the most likely responsible parties for the presence of the alleged defect. In general, the primary share of responsibility falls on the installing contractor unless the design is considered defective, in which case the designer would receive the primary share of blame. Regardless, fault allocation percentages must be evaluated on a case-by-case basis.

## VIII.   General Theory of Allocation Percentages

Attorneys often argue that allocation opinions are left for the trier of fact during binding dispute resolution, whether that be a judge, jury, or arbitrator(s). However, in certain technical cases, allocation opinions are sometimes allowed to assist the trier of fact. Regardless, allocation opinions can be valuable during the non-binding dispute resolution process, particularly when a large number of defendants and third-party defendants are involved in a matter because it provides a settlement roadmap during the non-binding dispute resolution process. The percentages that are often used for fault allocation are listed below by way of example.

**Example** – *Owner Rejects Work and the Contractor-Subcontractor Agreements Have Indemnification Clauses*
The dispute involves an owner claim of improper weather-resistive barrier (WRB) installation that is behind the stucco veneer. The general contractor hired a separate subcontractor to install the WRB. The stucco subcontractor covered up this patent defect. Per the general contractor's subcontract agreement forms that were used with the WRB and stucco subcontractors, the subcontractors must fully indemnify the general contractor for all defective work installed by the subcontractors. In addition, the subcontract agreements note that subcontractors are responsible for covering up patently defective work by others. In this instance, the general contractor might allocate fault as follows:

0% – General Contractor
25% – Stucco Subcontractor
75% – WRB Subcontractor

**Example** – *Owner Rejects Work and the Contractor-Subcontractor Agreements Have No Indemnification Clauses*
The dispute involves an owner-claim of improper flashing within the stucco veneer. The general contractor hired a separate flashing subcontractor for this work. The stucco subcontractor covered up this patent defect. Per the general contractor's subcontract agreement forms that were used with the stucco and flashing subcontractors, there is <u>no</u> indemnification provision listed where the subcontractors must fully indemnify the general contractor for all defective work installed by the subcontractors. In addition, the subcontract agreement forms note that

subcontractors are responsible for covering up patently defective work by others. In this instance, the reasonable fault allocation amongst the parties might be as follows:

10% – General Contractor
22.5% – Stucco Subcontractor (.9 x 25%)
67.5% – Flashing Subcontractor (.9 x 75%)

### Example – *Owner Rejects Work and the Contractor Notes It Is a Design Issue*

This dispute involves an owner-claim of a lack of control joints within the stucco veneer. The design does not detail any control joints on the building elevations on the architectural plans; however, the installation of control joints in stucco veneer is a building code requirement and the contractor and its stucco subcontractor are bound to building code requirements. The subcontract between the contractor and subcontractor has an indemnification provision that makes the subcontractor fully responsible for defective work that it installs. Here, the designer is required to design per code and the subcontractor is responsible for installing the stucco veneer per code. Because the building code provision that requires control joints in stucco veneers is a provision that a reasonable stucco subcontractor should know, so the designer and the subcontractor are held to be responsible here, and a mediator might propose the following fault allocation:

50% – Designer
50% – Stucco Subcontractor

# 14

# Conclusion

For AEC professionals, it is important to understand how to assert, evaluate, and defend against contract claims related to construction projects. This book provides claimants with a roadmap on how to properly assert contract claims as well as a roadmap for respondents on how to assess and rebut claims. Once one knows how to properly assert a claim, one knows how to evaluate and defend against a contract claim. The nine-step process outlined in this book goes from reviewing the dispute resolution procedures within the subject contract to binding dispute resolution, and it details everything in between from defining the type of dispute, fulfilling notice requirements, establishing entitlement, calculating delay durations, calculating damages, formatting and packaging the claim, and managing the non-binding dispute resolution process.

Contract termination claims are dealt with separately in this book due to the severe nature of this remedy. As discussed, the terminating party must be sure to closely follow the procedural requirements within the contract and be sure that a substantive basis exists to warrant a termination for cause. While the focus of this book is contract claims, the reader should be familiar with non-contract claims and defenses that are often included in pleadings for construction disputes. Understanding non-contract claims and defenses, at least at a cursory level, is important during settlement negotiations before binding dispute resolution is triggered, as it might impact a final holding during binding dispute resolution. Also, understanding a methodology for fault allocation is critical for asserting proper damages to multiple parties that might be partially responsible for a claimed contract issue.

In sum, effective dispute resolution administration is a great risk management tool for all parties associated with construction projects. The process outlined in this book will help recover funds that are owed to claimants and it can prevent payment of frivolous claims. Furthermore, it can prevent catastrophic results and promote reasonable settlements that avoid costly arbitration and litigation.

# Index